9-20-73

READINGS IN ECOLOGY

Edited by
Polley Ann Randolph
James Collier Randolph
Indiana University

MSS Information Corporation
655 Madison Avenue, New York, N.Y. 10021

This is a custom-made book of readings prepared for the courses taught by the editors, as well as for related courses and for college and university libraries. For information about our program, please write to:

MSS INFORMATION CORPORATION
655 Madison Avenue
New York, New York 10021

MSS wishes to express its appreciation to the authors of the articles in this collection for their cooperation in making their work available in this format.

Library of Congress Cataloging in Publication Data

Randolph, Polley Ann, comp.
   Readings in ecology.

   1. Ecology--Addresses, essays, lectures.
I. Randolph, James Collier, joint comp. II. Title.
QH541.145.R36   1973   574.5'08   72-13518
ISBN 0-8422-0291-9

Copyright © 1973
by
MSS INFORMATION CORPORATION
All Rights Reserved.

CONTENTS

Relationships Between Structure and Function
in the Ecosystem
    EUGENE P. ODUM.......................................5

The Energy Environment In Which We Live
    DAVID M. GATES......................................16

Energy Dynamics of a Food Chain of an
Old-Field Community
    FRANK B. GOLLEY....................................38

Community Metabolism in a Temperate Cold Spring
    JOHN M. TEAL.......................................86

Concluding Remarks
    G. EVELYN HUTCHINSON.............................130

Niche Segregation in Seven Species of Diplopods
    ROBERT V. O'NEILL................................143

The Influence of Interspecific Competition and
Other Factors on the Distribution of the
Barnacle *Chthamalus Stellatus*
    JOSEPH H. CONNELL...............................145

An Experimental Component Analysis of Population
Processes
    C.S. HOLLING....................................159

The Case for the Multispecies Ecological System,
with Special Reference to Succession and
Stability
    G. DENNIS COOKE, ROBERT J. BEYERS, and
    EUGENE P. ODUM..................................169

RELATIONSHIPS BETWEEN STRUCTURE AND FUNCTION IN THE ECOSYSTEM*

Eugene P. ODUM

I am honored and delighted to be here today to address the Ecological Society of Japan. I bring greeting from all members of the Ecological Society of America which will hold its next annual meeting at Oregon State University on the west coast of the U.S.A. After my visit to Japan I am expecting to have much to report to American ecologists about the progress of ecological research in Japan. We hope that in the years to come many of you will be able to visit the United States of America.

The topic I wish to discuss with you today is: Relationships between structure and function in the ecosystem. As you know ecology is often defined as: The study of interrelationships between organisms and environment. I feel that this conventional definition is not suitable; it is too vague and too broad. Personally, I prefer to define ecology as: The study of the structure and function of ecosystems. Or we might say in a less technical way: The study of structure and function of nature.

By structure we mean: (1) The composition of the biological community including species, numbers, biomass, life history and distribution in space of populations; (2) the quantity and distribution of the abiotic (non-living) materials such as nutrients, water, etc.; (3) the range, or gradient, of conditions of existence such as temperature, light, etc. Dividing ecological structure into these three division is, of course, arbitrary but I believe convenient for actual study of both aquatic and terrestrial situations.

By function we mean: (1) The rate of biological energy flow through the ecosystem, that is, the rates of production and the rates of respiration of the populations and the community; (2) the rate of material or nutrient cycling, that is, the biogeochemical cycles; (3) biological or ecological regulation including both regulation of organisms by environment (as, for example, in photoperiodism) and regulation of environment by organisms (as, for example, in nitrogen fixation by microorganisms). Again, diving ecological function into these three division is arbitrary but convenient for study.

Until recently ecologists have been largely concerned with structure, or what we might call the descriptive approach. They were content to describe the conditions of existence and the standing crop of organisms and materials. In recent years equal emphasis is being placed on the functional approach as indicated by the increasing number of studies on productivity and biological regulation. Also the use of experimental methods, both in the field and in the laboratory, has increased. Today, there exists a very serious gap between the descriptive and the functional approach. It is very important that we bring together these two schools of ecology. I should like to present some suggestions for bridging this gap.

The main features of the structure of a terrestrial and an aquatic ecosystem may be illustrated by comparing an open water community, such as might be found at sea or in a large lake, with a land community such as a forest. In our discussion we shall consider these two types as models for the extremes in a gradient of communities which occur in our biosphere. Thus, such ecosystems as estuaries, marshes, shallow lakes, grasslands and agricultural croplands will have a community structure intermediate between the open water and forest types.

Both aquatic and terrestrial community types have several structural features in common. Both must have the same three necessary biological components: (1) Producers or green plants capable of fixing light energy (i.e., autotrophs); (2) animals or macro-consumers which consume particulate organic matter (i.e., phagotrophs); and (3) microorganism decomposers which dissolve

---

* Address given at the 9th Annual Meeting of the Ecological Society of Japan, April 4, 1962

organic matter releasing nutrients (i.e., osmotrophs). Both ecosystems must be supplied with the same vital materials such as nitrogen, phosphorus, trace minerals, etc. Both ecosystems are regulated and limited by the same conditions of existence such as light and temperature. Finally, the arrangement of biological units in vertical space is basically the same in the two contrasting types of ecosystems. Both have two strata, an autotrophic stratum above and a heterotrophic stratum below. The photosynthetic machinery is concentrated in the upper stratum or photic zone where light is available, while the consumer-nutrient regenerating machinery is concentrated largely below the photic zone. It is important to emphasize that while the vertical extent or thickness of communities varies greatly (especially in water), light energy comes into the ecosystem on a horizontal surface basis which is everywhere the same. Thus, different ecosystems should be compared on a square meter basis, not on a cubic or volume basis.

On the other hand, aquatic and terrestrial ecosystems differ in structure in several important ways. Species composition is, of course, completely different; the roles of producers, consumers and decomposers are carried out by taxonomically different organisms which have become adapted through evolution. Trophic structure also differs in that land plants tend to be large in size but few in number while the autotrophs of open water ecosystems (i.e., phytoplankton) are small in size but very numerous. In general, autotrophic biomass is much greater than heterotrophic biomass on land, while the reverse is often true in the sea. Perhaps the most important difference is the following: The matrix, or supporting framework, of the community is largely physical in aquatic ecosystems, but more strongly biological on land. That is to say, the community itself is important as a habitat on land, but not so important in water.

Now, we may ask: How do these similarities and differences in structure affect ecological function?

One important aspect of function is shown in Fig. 1 which compares energy flow in an aquatic and a terrestrial ecosystem. The upper diagram is an energy flow model for a marine community; the lower diagram is a comparable model for a forest. The boxes represent the average standing crop biomass of organisms to be expected; the light gray boxes are the autotrophs, the darker boxes are the heterotrophs. Three trophic levels are shown: (1) Producers, the phytoplankton of the sea and the leaves of the forest trees; (2) primary consumers (herbivores, etc.); and (3) secondary consumers (carnivores). The pipes or flow channels represent the energy flow through the ecosystems beginning with the incoming solar energy and passing through the successive trophic levels. At each transfer a large part of the energy is dissipated in respiration and passes out of the system as heat. The amount of energy remaining after three steps is so small that it can be ignored in so far as the energetics of the community are concerned. However, tertiary consumers ("top carnivores") can be important as regulators; that is, predation may have an important effect on energy flow at the herbivore level.

All numbers in the diagrams are in terms of large or Kilogram Calories and square meters; standing crop is in terms of KCal. /M^2; energy flow is in terms of KCal./M^2/day. The diagrams are drawn so that the area of the boxes and the pipes are proportional to the magnitude of the standing crops and energy flows respectively. The quantities shown are a composite of measurements obtained in several different studies; some of the figures for higher trophic levels are hypothetical since complete information is not yet available for any one ecosystem. The marine community is particularly based on the work of Gorden RILEY (Long Island Sound) and H.W. HARVEY (English Cannel), and the forest on the work of J.D. OVINGTON (pine forest) and unpublished data on terrestrial communities from our research group at the University of Georgia.

The autotrophic-heterotrophic stratification, which we emphasized as a universal feature of community structure, results in two basic food chains as shown in both diagrams (Fig. 1). The consumption of living plants by herbivores which live in the autotrophic stratum together with their predators may be considered as the *grazing food chain*. This is the classical food chain of ecology, as, for

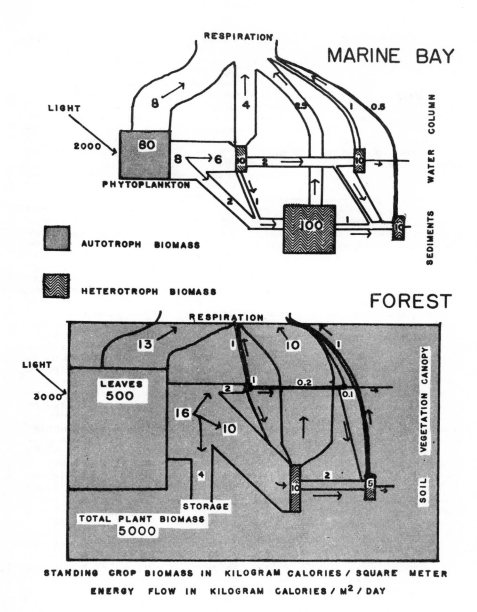

Fig. 1. Energy flow models for two contrasting types of ecosystems, an open water marine ecosystem (upper diagram) and a terrestrial forest (lower diagram)

Standing crop biomass(in terms of KCaL/M²) and trophic structure are shown by means of shaded rectangles. Energy flows in terms of KCal/M²/day (average annual rate) are shown by means of the unshaded flow channels. The aquatic system is characterized by a small biomass structure (hence the habitat is largely physical) while the forest has a very large biomass structure (hence the habitat is strongly biological). In both types of systems the energy of net primary production passes along two major pathways or food chains: (1) the grazing food chain (upper sequence in the water column or vegetation), and (2) the detritus food chain (lower sequence in sediments or soil).

The marine diagram is based on work of RILEY and HARVEY, the forest diagram on the work of OVINGTON and unpublished data from research at the University of Georgia. In some cases figures are hypothetical since no complete study has yet been made of any ecosystem. Hence, the diagrams should be considered as "working models" which do not represent any one situation.

example, the phytoplankton-zooplankton-fish sequence or the grass-rabbit-fox sequence. However, a large proportion of the net production may not be consumed until dead, thus becoming the start of a rather different energy flow which we may conveniently designate as the *detritus food chain*. This energy flow takes place largely in the heterotrophic stratum. As shown in Fig. 1 the detritus energy flow takes place chiefly in the sediments of water systems, and in the litter and soil of land systems.

Ecologists have too often overlooked the fact that the detritus food chain is the more important energy pathway in many ecosystems. As shown in Fig. 1 a larger portion of net production is estimated to be consumed by grazers in the marine bay than in the forest; nine-tenths of the net production of the forest is estimated to be consumed as detritus (dead leaves, wood, etc.). It is not clear whether this difference is a direct or indirect result of the difference in community structure. One tentative generalization might be proposed as follows: communities of small, rapidly growing producers such as phytoplankton or grass can tolerate heavier grazing pressure than communities of large, slow-growing plants such as trees or large seaweeds.

Grazing is one of the most important practical problems facing mankind; yet we know very little about the situation in natural ecosystems. Well-ordered and stable ecosystems seem to have numerous mechanisms which prevent excessive grazing of the living plants. Sometimes, predators appear to provide the chief regulation; sometimes weather or life history characteristics (limited generation time or limited number of generations of herbivores) appear to exercise control. Unfortunately, man with his cattle, sheep and goats often fails to provide such regulation with result that overgrazing and declining productivity is apparent in large areas of the world, especially in grasslands. A study of the division of energy flow between grazing and detritus pathways in stable natural ecosystems can provide a guide for man's utilization of grasslands, forests, the sea, etc.

The energy flow diagrams, as shown in Fig. 1, reemphasize the difference in biomass as mentioned previously. Autotrophic biomass is very large and envelops or encloses the whole community in the forest; such extensive biological structure buffers and modifies physical factors such as temperature and moisture. In contrast, the aquatic community stands naked or exposed to the direct action of physical factors. In the marine situation the animal biomass often exceed the plant biomass, and sessile animals (oysters, barnacles, etc.) instead of plants often provide some protection or habitat for other organisms.

Despite the large difference in relative size of standing crops in the two extreme types of ecosystems, the actual energy flow may be of the same order of magnitude if light and available nutrients are similar. In Fig. 1 we have shown the available light (absorbed light) and the resulting net production as being somewhat lower in the marine community, but this may not always be true. Thus, 80 KCals of phytoplankton may have a net production almost as large as 5000 KCals of trees (or 500 KCals of green leaves). Therefore, productivity is not proportional to the size of the standing crop except in special cases involving annual plants (as in some agriculture). Unfortunately, many ecologists confuse productivity and standing crop. The relation between structure and function in this case depends on the size and rate of metabolism (and rate of turnover) of the organisms.

To summarize, we see that biological structure influences the pattern of energy flow, particularly the fate of net production and the relative importance of grazers and detritus consumers. However, total energy flow is less affected by structure, and is thus less variable than standing crop. A functional homeostasis has been evolved in nature despite the wide range in species structure and in biomass structure.

So far we have dealt with structure in relation to one aspect of function of the entire ecosystem. Now let us turn to structure and function at the population level and consider a second major aspect of function, namely, the cycling of nutrients. As an example I shall review the work of Dr. Edward J. KUENZLER at the University of Georgia Marine Institute on Sapelo Island. The study concerned a species of mussel of the genus *Modiolus* in the intertidal salt marshes. There are similar species of filter-feeding mollusca

ROLE OF A MOLLUSCAN POPULATION IN NUTRIENT CYCLING AND ENERGY FLOW IN A SALT MARSH ECOSYSTEM
SAPELO ISLAND GEORGIA

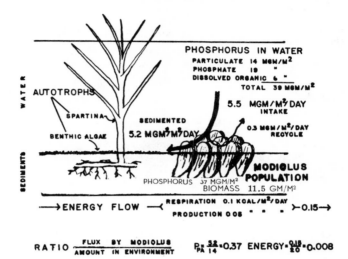

Fig. 2. The effect of a population of mussels (*Modiolu*s) on energy flow and the cycling of phosphorus in a salt marsh ecosystem according to the study of Dr. E.J. KUENZLER at the University of Georgia Marine Institute, Sapelo Island, Georgia, U.S.A.

From the standpoint of the ecosystem as a whole the population has a much greater effect on the cycling of phosphorus than on the transformation of energy. The study illustrates one often overlooked function of animals, that of nutrient regeneration. See text for details of the study.

in the intertidal zone in all parts of the northern hemisphere.

First, we shall take a look at the salt marsh ecosystem and the distribution of the species in the marsh. The mussels live partly buried in the sediments and attached to the stems and rhyzomes of the marsh grass, *Spartina alterniflora*. Individuals are grouped into colonies (clumped distribution), but the colonies are widely scattered over the marsh. Numbers average 8/M² for the entire marsh and 32/M² in the most favorable parts of the marsh. Biomass in terms of ash-free dry weight averages 11.5 gms/M². When the tide covers the colonies the valves partly open and the animals begin to pump large quantities of water.

Fig. 2 illustrates the role of the mussel population in phosphorus cycling and energy flow according to Dr. KUENZLER's data. Each day the population removes a large part of the phosphorus from the water, especially the particulate fraction. Most of this does not actually pass through the body but is sedimented in the form of pseudofeces which fall on the sediments. Thus, the mussel make large quantities of phosphorus available to microorganisms and to the autotrophs (benthic algae and marsh grass). As shown along the bottom of the diagram (Fig. 2) the energy flow was estimated to be about 0.15KCals/M² /day.

The most important finding of the study is summarized in the bottom line below the diagram (Fig. 2) which shows the ratio between flux and amount. Note that over one third of the 14 mgms of particulate phosphorus is removed from the water each day by the population, and there is retained in the marsh. In contrast, less than one per cent of the 20 KCals of potential energy (net production estimate) available is actually utilized by the mussel population. In other words, the mussel population has a much more important effect on the community phosphorus cycle than it has on community energy flow. Or one might say that the role of the mussel in conserving nutrients in the ecosystem is more important than its role as energy transformer. In other words, the mussel population would be of comparatively little importance as food for man or animals (since population growth or production is small), but is of great importance in maintaining high primary production of the ecosystem.

To summarize, the mussel study brings out two important points: (1) It is necessary to study both energy flow and biogeochemical cycles to determine the role of a particular species in its ecosystem, (2) animals may be important in the ecosystem not only in terms of food energy, but as agents which make basic nutrient more available to autotrophs.

Finally, I think it is highly significant that the most productive ecosystems of the biosphere are those in which autotrophic and heterotrophic strata lie close together, thus insuring efficient nutrient regeneration and recycling. Estuaries, marshes, coral reefs and rice fields are examples of such productive ecosystems.

Now let us consider the third important

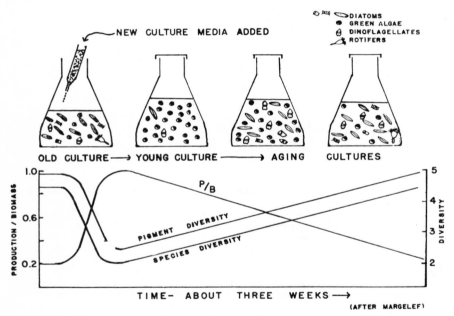

Fig. 3. The MARGELEF model of ecological succession showing a simple type of succession which can be demonstrated in laboratory cultures

The flasks show changes in species composition occurring when succession is set in motion by the introduction of new nutrient media into an old "climax" culture. The gragh shows resultant changes in two aspects of diversity and in the relation between production and biomass (P/B). See text for details of the experiment.

aspect of ecological function, that is, community regulation. Ecological succession is one of the most important processes which result from the community modifying the environment. Fig. 3 illustrates a very simple type of ecological succession which can be demonstrated in a laboratory experiment. Yet the basic pattern shown here is the same as occurs in more complex succession of natural communities. The diagram (Fig. 3) was suggested to me by Dr. Ramon MARGELEF, hence we may call it the MARGELEF model of succession.

At the top of the diagram (Fig. 3) are a series of culture flasks containing plankton communities in different stages of succession. The graph shows changes in two aspects of structure and in one aspect of function. The first flask on the left contains an old and relatively stable community; this flask represents the climax. Diversity of species is high in the climax; species of diatoms, green flagellates, dinoflagellates and rotifers are shown in the diagram to illustrate the variety of plants and animals present. Biochemical diversity is also high as indicated by the ratio of yellow plant pigments (optical density at $430m\mu$) to chlorophyll-a (optical density at $665m\mu$). On the other hand the ratio of production to biomass (P/B in Fig. 3) is low in the old or climax culture, and gross production tends to equal community respiration. If we add fresh culture medium to the old culture, as shown in Fig. 3, ecological succession is set in motion. An early stage in succession is shown in the second flask. Species diversity is low, with one or two species of phytoplankton dominant. Chlorophylls predominate so that the yellow/green ratio (O.D. 430 / O.D. 665) is low, indicating low biochemical diversity. On the other hand, production now exceed respiration so that the ratio of production to biomass becomes higher. In other words, autotrophy greatly exceed heterotrophy in the pioneer or early succession stage. The two flasks on the right side of the diagram (Fig. 3) show the gradual return to the climax or steady state where autotrophy tends to balance heterotrophy.

The changes which we have just described are apparently typical of all succession regardless of environment or type of ecosystem. Although much more study is needed, it appears that differences in community structure mainly affect the time required, that is, whether the horizontal scale (X-axis in Fig. 3) is measured in weeks, months or years. Thus, in open water ecosystems, as in cultures, the community is able to modify the physical environment to only a small extent. Consequently, succession in such ecosystems is brief, lasting perhaps for only a few weeks. In a typical marine pond or shallow marine bay a brief succession from diatoms to dinoflagellates occurs each season, or perhaps several times each season. Aquatic ecosystems characterized by strong currents or other physical forces may exhibit no ecological succession at all, since the community is not able to modify the physical environment. Changes observed in such ecosystems are the direct result of physical forces, and are not the result of biological processes; consequently, such changes are not to be classed as ecological succession.

In a forest ecosystem, on the other hand, a large biomass accumulates with time, which means that the community continues to change in species composition and continues to regulate and buffer the physical environment to a greater and greater degree. Let us refer again to Fig. 1 which compares a forest with an aquatic ecosystem. The very large biological structure of the forest enables the community to buffer the physical environment and to change the substrate and the microclimate to a greater extent than is possible in the marine community.

Recent studies on primary succession on such sites as sand dunes or recent volcanic lava flows indicate that at least 1000 years may be required for development of the climax. Secondary succession on cut-over forest land or abandoned agricultural land is more rapid, but at least 200 years may be required for development of the stable climax community. When the climate is severe as, for example, in deserts, grasslands or tundras the duration of ecological succession is short since the community can not modify the harsh physical environment to a very large extent.

To summarize, I am suggesting that the basic pattern of functional change in ecological succession is the same in all ecosystems, but that the species composition, rate of change and duration of succession is determined by the physical environment and the resultant community structure.

The principles of ecological succession are of the greatest importance to mankind. Man must have early successional stages as a source of food since he must have a large net primary production to harvest; in the climax community production is mostly consumed by respiration (plant and animal) so that net community production in an annual cycle may be zero. On the other hand, the stability of the climax and its ability to buffer and control physical forces (such as water and temperature) are desirable characteristics from the viewpoint of human population. The only way man can have both a productive and a stable environment is to insure that a good mixture of early and mature successional stages (i.e. "young nature" and "old nature") are maintained with interchanges of energy and materials. Excess food produced in young communities helps feed older stages which in return supply regenerated nutrients and help buffer the extremes of weather (storms, floods, etc.).

In the most stable and productive of natural situations we usually find such a combination of successional stages. For example, in continential shelf marine areas such as the inland sea of Japan the young communities of plankton feed the older, more stable communities on the rocks and on the bottom (i.e., benthic communities). The large biomass structure and diversity of the benthic communities provide not only habitat and shelter for life history stages of pelagic forms, but also provide regenerated nutrients necessary for continued productivity of the plankton.

A similar favorable situation exists in the Japanese terrestrial landscape where productive rice fields on the plains are intermingled with diverse forests on the hills and mountains. The rice fields, of course, are, ecologically speaking, "young nature" or early successional communities with very high rates of net community production which are maintained as such by the constant labor of the farmer and his machines. The forests represent older, more diverse and self-sustaining communities which have a lower net production, but do not require the constant attention of man. It is important that both ecosystems be considered together in proper relation. If the forests are destroyed merely for the sake of temporary gain in wood production, then the water and soil will wash down from the slopes and destroy the productivity of the plains. In my brief travels in Japan I have noted an unfortunate tendency in some areas to consider only the productive aspect of forests, and consequently to ignore their protective value. Complete deforestation of slopes may yield more wood for the time being but is ecologically a very dangerous procedure; also rebuilding the ecosystem is always more expensive than maintaining it in good condition. I believe ecologists should be more aggressive in bring these principles to the attention of those charged with responsibility of national resources. Especially, ecologists need to provide good data which demonstrates the value of forests and other mature-type ecosystems in maintaining water and nutrient cycles. The value of forest should not be measured only in terms of net production.

My purpose in reviewing the three basic aspects of ecological function (that is, energy flow, nutrient cycles, and biological regulation) is to emphasize that we must study both structure and function if we are really to understand and control nature. Usually the study of function is more difficult than the study of structure; hence functional ecology has lagged behind the descriptive ecology. To study function we must measure the rate of change per unit of time, and not just the situation at any one time. That is, we must measure the rate of energy flow, not just the standing crop; we must measure the rate of exchange of phosphorus, not just the amount present in the ecosystem; and we must measure the degree of regulation, not just describe it. I should like to close my lecture with a brief discussion of how new technics resulting from the peaceful uses of atomic energy may help us to make better measurements of ecological function. I refer, of course, to the use of radioactive tracers in ecological research.

We should first emphasize two points about radioactive tracers: (1) Tracer technics do not solve any problems alone, but may be useful in conjection with other methods. Tracers, do extend our powers of observation greatly. Just as the microscope extended our powers of observation of biological structure, so tracers have extended our powers of

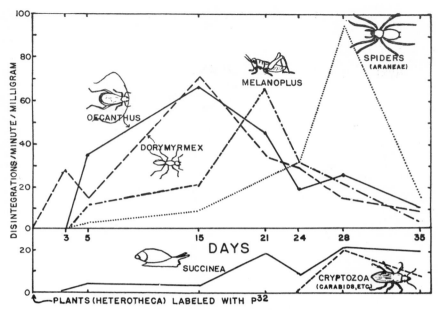

Fig. 4. The results of a simple experiment in which a radioactive tracer (P^{32}) was used to isolate a food chain in an intact "old-field" (i.e. early stage of succession on abandoned agricultural land) community

The buildup of P^{32} activity in the biomass of 6 major invertebrate populations following the labeling of a single species of dominant plant is shown in the diagram. This tracer technic resulted in the separation of certain trophic and habitat niches in the intact and undisturbed community. See text for details of the experiment.

observation of function. (2) The amount of radioactivity employed in a tracer experiment can be so small that there need be no hazard to the investigator, and no effect on organisms or environment. Instruments are now so sensitive that amounts of radioactivity far less than that contained in a radium-dial watch can be easily detected.

In the intact ecosystem it is often difficult to determine the exact energy source or food being utilized by various organisms. Likewise, it is difficult to measure the rate of metabolism of free living populations. radioactive tracers can aid in both of these important determinations in plants, animals and microorganisms.

Fig. 4 shows the results of a simple experiment in which all individuals of one species of plant within a large quadrat in a natural grassland community were labeled with radioactive phosphorus (P^{32}). This was done by placing a small drop of P^{32} (in form of phosphate) on a leaf of each individual plant of the species in question; within a short time the labeled phosphate spreads to all parts of the plant. In the experiment shown in Fig. 4 one of the dominant plants, *Keterotheca subaxilaris*, was labeled, while all other plants in the quadrat remained unlabeled. During the 30 days following the labeling samples of the invertebrates were collected and the concentration of P^{32} in their tissues determined. According to this procedure, any animal which becomes radioactive must have fed upon the plant, or must have eaten an animal which had previously fed upon the plant. Such a procedure enables us to isolate a single food chain (that is, a food chain beginning with a single species of autotroph) in an intact community.

In Fig. 4 the buildup of P^{32} radioactivity in six major populations is shown. Radioactivity in terms of disintegrations per minute per milligram of live weight is indicated on the Y-axis, and time is indicated on the X-axis. A small ant (Formicidae) and a small cricket (Orthoptera) were the first animals to reach a peak in radioactivity; these species were the most active animals as indicated by general observation. Larger plant feeders such as grasshoppers (*Melanoplus*) reached a peak at a later time, while predators such as spiders did not reach maximum levels until about four weeks after the labeling of the plant. There was also a delay in appearance of P^{32} in animals living on the ground or in the surface litter (carabids, snails, etc.).

As shown in Fig. 4 plotting radioactivity in the biomass against time resulted in a good separation of certain trophic and habitat niches. The experiment indicated that the ant, *Dorymyrmex*, was functioning as a herbivore rather than a predator since rapid buildup of radioactivity would not be expected if the species was feeding on other insects only. Since aphids could not be found on the plants it was tentatively concluded that the ants were feeding directly on plant juices at the time of the experiment (which was in the late spring of the year).

We belive that procedures involving the labeling of single energy sources in intact communities can be very useful not only in determining food intake and trophic position of free-living populations, but also perhaps in determining the rate of feeding. The more rapid the feeding the sooner the transfer of the tracer will be observed.

As the final example I would like to mention very briefly the work which. Mr. Jiro MISHIMA did while he was studying with us in the U.S.A. During the past year our group at the University of Georgia has been interested in investigating the possibility that the

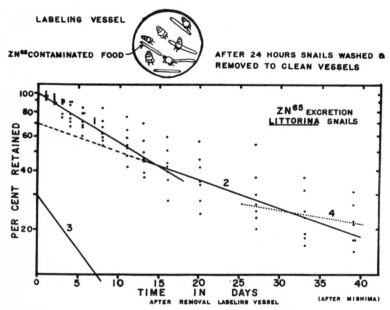

Fig. 5. The procedure used by Mr. Jiro MISHIMA in his study of the excretion (biological half-life) of the tracer, Zinc-65, by *Littorina*

The slope of regression line number 2 was found to be significantly affected by air temperature and body size in a manner parallel to expected oxygen consumption (i.e. rate of metabolism). See text for possible significance of this type of study.

excretion rate of isotopes may be used as indices of function. Theoretically, an active population should excrete a tracer more rapidly than an inactive group of organisms. If a good relation between excretion rate and activity can be demonstrated, then we would be able to estimate activity rate in wild animals provided only that we can recapture individuals or samples of the population after the labeled individuals had been released in their natural habitat. Preliminary trial experiments have now been completed in which the excretion rate of several radioactive tracers has been measured in insects, Crustacea, mollusks and fish under different environmental conditions.

Mr. MISHIMA studied the excretion rate of Zinc-65 in the salt marsh snail, *Littorina irrorata*, in relation to air temperature and body size. The general procedure is shown in Fig. 5. Snails freshly collected from the marsh were placed in a grass "labeling bowl" as shown at the top of the figure. A solution of $Zn^{65}Cl$ was placed on food in the bowl. After 24 hours the snails usually ingested a small amount of radioactive zinc. The snails were then washed thoroughly to remove surface contamination and placed in non-radioactive environments either under laboratory or field conditions. At intervals of one to three days marked individuals were placed in a well scintillation counter to determine how much of the tracer remained in the body. In the laboratory experiments the snails were transferred to clean bowls after each determination in order to avoid the possibility of re-ingestion of excreted material.

As shown in Fig. 5 the Zn^{65} tracer tended to be eliminated in two phases. Excretion was rapid during the first 10 days, and less rapid for the next 20 days or so. It is the second regression line (line 2 in Fig. 5) which is of greatest ecological interest since it shows the rate of loss of the tracer which was more completely assimilated into the organic biomass. Regressions were calculated for each individial by IBM electronic computer from logarythms of successive counts (corrected for radiological decay), and the slope of the line converted to a biological half-life figure. By biological half-life we mean: The time in days required for one-half of the tracer to be eliminated. The biological half-life of three sizes of snails (small, medium and large) were determined under four conditions: laboratory bowls at 10°, 25° and 30° and field conditions (natural marsh habitat) where temperature fluctuated between 9° and 31° but average 19°C.

Since the details of this study are to be published soon only a brief mention of the results will be needed. Analysis of variance of the data showed that both air temperature and body size had a highly significant effect of biological half-life of the assimilated tracer. The larger the snail, or the lower the temperature, the slower was the excretion rate (i.e. the longer the biological half-life). Furthermore, excretion rate as related to body size and temperature paralleled oxygen consumption rates as determined in a previous study on this species. Especially interesting was the fact that biological half-life in the field at 19°C was shorter (hence excretion more rapid) than in the laboratory at 25° or 30°C, suggesting that snails in the field were more active than when confined to laboratory bowls.

These preliminary results indicate that Zn^{65} may be useful as an index to the rate of metabolism. However, much work need to be done before such an "atomic meter" can be perfected. Probably a mixture of tracers will eventually prove useful, each tracer measuring a different aspect of function.

If we can discover how to measure the rate or activity or metabolism in free-living individuals of populations in intact ecosystems, then we will have taken a giant step towards better understanding of the relation of function to structure at the ecological level.

THE ENERGY ENVIRONMENT IN WHICH WE LIVE

By DAVID M. GATES

MY FINE radiating, convecting, sweating friends; you are sitting in a pure infrared thermal environment, you are inside a blackbody cavity which is only slightly modified by diffuse visible light. Man has been clever in erecting a barrier between himself and the vicissitudes of the weather and, in fact, controlling his captive climate. Other animals must seek natural nests and local lairs for comfort or else migrate with the changing climate. Plants are rooted to their environment and cannot walk out of the searing sun or escape the wind-swept plains or mountain slopes.

The planet Earth is indeed unique among the planets in our solar system, neither too close to the sun nor too far from the sun, and rotating at a rate so as to produce night and day in close succession. Life has evolved over the surface of the planet, protected from the cosmic cold of outer space and from the ionizing radiations of the sun by an envelope of atmosphere, and shielded from cosmic particles and solar storms by a magnetic field. How life evolved is still a matter of considerable conjecture, but life in its present form and distribution is amenable to close scrutiny and detailed understanding by man. How are plants and animals limited in their geographical distributions, what environmental factors control their evolution, how rapidly do they evolve, how do they interact to form ecosystems? These are etiological questions which challenge modern science to the full depth and breadth of its scope involving the laws of physics, chemistry, and biology. There is no short and easy road to the answers and certainly no single branch of science alone will lead to success. Organic evolution and organismic behavior is the consequence of physical laws which, by themselves, are well understood. The interaction of these physical laws with organic forms is enormously complex and diversified. To understand this interaction, one must cope with the physical and chemical laws, but with a full appreciation for the biological responses and involvement. The study of relationships of organisms to their environment is known as ecology.

Most organisms live within a rather narrow temperature range, with a maximum temperature becoming more disastrous than a minimum temperature. As evident from Figure 1, most biological activity is confined to the temperature range 0°C. to 60°C., a rather narrow temperature range when one considers that absolute zero is 273°C. below 0°C. Al-

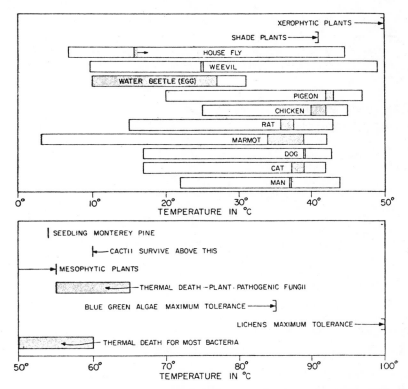

FIG. 1. The tolerable temperature regimes for many plants and animals. Shaded areas represent the normal temperature for the animal and the unshaded areas represent the extreme tolerances. Information is from the *Handbook of Biological Data*, edited by William S. Spector (1956), W. B. Saunders Co. Publishers, Philadelphia, Pennsylvania.

though many simple organisms and a few higher forms can remain viable after exposure to absolute zero, most plants and animals do not carry on biological activity below 0°C. Some of the higher animals maintain a very narrow range in body temperature through complex physiological controls. Hence it becomes evident, for both plants and animals, that the temperature of an organism resulting from environmental influence is of critical importance. It is in this vein that the current discussion concerning the nature of environments is conceived.

Energy Flow

The surface of the Earth has been described in terms of various climates representing deserts, grass lands, tropical forests, boreal forests, alpine tundra, arctic tundra, island atolls, etc. But what does climate mean, or for whom or what purpose is climate described? The dictionary gives two definitions of climate. The first pertaining to weather conditions and the second as "any prevailing conditions affecting life." It is this second definition which must be used precisely in ecology, whereas far too often the first definition has been used. One must describe the

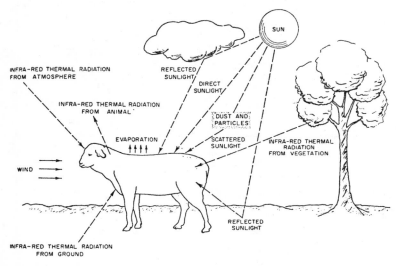

FIG. 2. The streams of energy flow to and from an organism in its natural environment. Figure from Gates (1962).

environment in terms of the flow of energy to the organism and not solely in terms of meteorological parameters. It is not the temperature of the walls of this room which is of importance to you, but rather the radiant heat from the walls received by your body which has meaning. And so it is for the environment of plants and animals that the flow of energy is meaningful and not simply the meteorological parameters describing the properties of the atmosphere. The meteorological parameters are useful for they enter into the energy flow relationships, but there are certain constants which couple the meteorological parameters to the organism through the energy flow relationships. Furthermore, the meteorological parameters enter into the energy flow relationships in a complex manner.

In Figure 2 the complexity of the energy exchange between an organism and its environment is given in diagrammatic form. Here are shown the streams of direct, scattered, and reflected sunlight, infrared thermal radiation from the atmosphere, from the ground, and from surrounding objects such as the vegetation; the reradiation of infrared radiation by the animal; convection by wind, which may either cool or warm the animal according to whether the wind is cooler or warmer than the surface temperature of the animal; and the evaporation of moisture or sweating, which cools the animal and which, at times of severe heat stress, can save the animal's life. All of these terms may be represented by means of the following equation:

$$(1 + r)(S + s) + R + LE + G + C + P + M = 0 \qquad (1)$$

where

$(1 + r)$
$(S + s)$ = direct plus scattered and reflected sunlight.
R = infrared thermal radiation from atmosphere, ground and surrounding objects and the reradiation from the organism.
LE = evaporation of transpiration with L being the latent heat of evaporation and E the quantity of moisture lost.
G = conduction of heat to or from the organism; which applies mainly for objects immersed in soil, snow, etc.
C = convection of heat to or from the organism including free convection in still air or water, and forced convection in windy air or moving water.
P = photosynthesis.
M = metabolism.

Solar Radiation

The solar radiation term is strongly variable over the globe varying with the time of day, season, cloud conditions, dust, and with the nature of the precise location depending upon the degree of exposure. The seasonal variation of sunlight is enormous in polar regions and only slight in equatorial regions. The diurnal changes from night to day are dramatic in the tropics where, according to Rudyard Kipling, "the dawn comes up like thunder"; but polar regions are described in the words of Robert Service "into the bowl of the midnight sky, violet, amber, and rose." The distribution of sunlight over the earth and its seasonal and diurnal variation is well known and is thoroughly discussed in many books on climatology and by Gates (1962).

The spectral distribution of solar radiation and skylight is of considerable interest when one is considering biological adaptation and behavior. The spectral distribution of solar radiation plus skylight is shown for a clear day at sea level in Figure 3. Here the abscissa is given in wave numbers (cm^{-1}) defined as the reciprocal of the wave length given in centimeters. A similar plot is given in Figure 4 where a wavelength scale is given above the curve. A frequency plot is used here since, according to Planck's law, energy is proportional to the frequency. In this way, equal areas under the curve represent equal energy. With the usual linear wave-length plot of solar energy one gets the misimpression that the peak of the solar energy distribution is in the green of the visible, whereas the energy distribution peaks in the near infrared. It is interesting to note in Figure 3 how the skylight is strongly distributed towards the high frequency, short wave length, blue end of the visible spectrum. This is as it should be since the sky is blue due to Rayleigh scattering. The low frequency, long wave length, infrared end of the spectrum has the solar radiation strongly depleted by the atmospheric water vapor and carbon dioxide absorptions. The high frequency, or short wave length, ultraviolet termination of the solar energy received

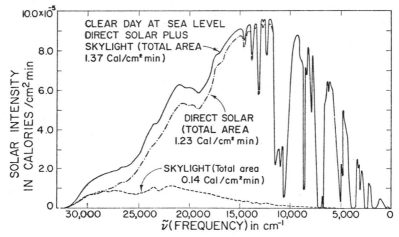

FIG. 3. Spectral distribution of direct sunlight, scattered skylight, and the sum of the two on a horizontal surface at sea level for a typical clear day as a function of the frequency of the light.

at the Earth's surface is abruptly ended through absorption by ozone and oxygen molecules resident in the Earth's upper atmosphere.

Figure 4 shows what a small segment of the total solar energy received at the Earth's surface is perceived by the human eye. The luminosity curve for the human eye represents the response by the rods and cones of the retina to each frequency of incident light. The normal eye receiving the solar energy given in Figure 4 would see this energy as equivalent to 8600 foot candles and the luminosity curve is given in terms of the scale on the right hand side of Figure 4. Most barrier-layer photo-cells have spectral response quite similar to that of the human eye and, when used in various forms of light meters, photometers, or illuminometers, are calibrated in luminosity units, foot candles. Because of the availability of this type of instrument, many measurements of the light intensity in the field have been made in foot candles. This is most unfortunate indeed, since a measure of the light intensity in foot candles has very little relationship to the total radiant flux of daylight. Enormous changes in the spectral distribution of the incident light can occur without substantial changes occurring in the luminous flux. For problems of illumination where the human eye is involved, measurements in foot candles are useful, but for all other biological systems it is far better to measure the radiant flux in either calories or ergs/cm²/minute.

Plants on the Earth have adapted in a most interesting manner to the frequency distribution of solar radiation. They possess a high absorp-

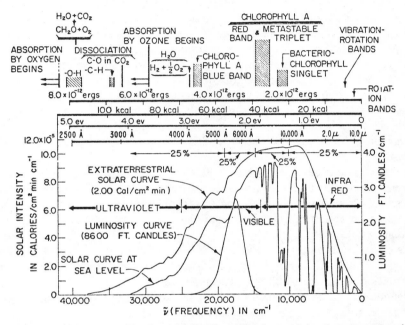

FIG. 4. The frequency distribution of solar radiation and its interpretation in terms of biologically significant reactions. Since energy per quantum is proportional to frequency, various energy scales are shown horizontally and a wavelength scale is given for general orientation. The visual response of the human eye is given in order to show the narrow visible range.

tivity at short wave lengths and throughout the visible, where they must absorb solar radiation efficiently for photosynthesis. Immediately to the long wave-length side of the red chlorophyll band, at a wave length of 7000 Å, the absorptivity drops precipitously from above 90% to less than 20%. This also shows up in terms of the high reflectivity of vegetation at photographic infrared wave lengths. This low absorptivity in the near infrared, just beyond the red, is valuable to a plant since the solar radiation reaches its maximum intensity at these frequencies. The low absorptivity reduces the heat absorbed by a plant and plays a critical role in keeping a plant from getting too warm in the midday sun. Further out in the infrared, at about a wave length of 2 microns, the absorptivity reaches 95%. The intensity of solar radiation is dropping sharply at 2 microns and therefore it does not overheat the plant to absorb strongly beyond this wave length. However, the high absorptivity means a high emissivity at the longer wave lengths so that the plant cools radiatively with efficiency. This mechanism is also critical for a plant to keep cool under radiation from the noonday sun.

Figure 4 displays two types of energy relationships. The ordinate represents the intensity or flux of the radiation in calories/cm²/min/cm^{-1}. Each quantum of radiation of a particular frequency, ν, also represents energy according to Planck, since $E = h\nu$, where h is Planck's constant. The equivalent energy per mole quanta corresponding to each frequency of radiation is shown on three horizontal scales corresponding to the frequency scale at the bottom of the figure. It should be noticed that the energy scales are strictly linear. These energy scales are given in ergs/mole, kcal/mole, and eV/mole so that the energies involved in certain physical or chemical phenomena can be compared. The corresponding wave-length scale is also shown and it should be noticed that it is nonlinear.

Above the energy scales are shown a few of the photochemical reactions which are of importance to biological systems. It should be noticed that absorption by ozone in the stratosphere shields the organic complexes near the Earth's surface from dissociation by ultraviolet radiation which would break C–O, C–H, O–H bonds, and others. It is not the place here to discuss in any detail the phenomena of photosynthesis and the origin of life. Rather it is appropriate to point out the frequencies of radiation which can act constructively to form complex organic compounds and those frequencies which will act destructively to break down organic molecules. It is this rather sharp distinction between the constructive frequencies reaching the surface of the planet and the screening by the atmosphere of the destructive frequencies, coupled with suitable heat load or temperature regime at the surface, which has permitted higher forms of life to evolve here on Earth. When considering the possibility of life on other planets one must make reasonable estimates of the radiation spectrum and heat load relationships at the surface in order to be assured that the constructive radiations are present and the destructive radiations are not present. The spectral characteristics of radiation at the surface will depend directly on the molecular composition of the planetary atmosphere.

Infrared Thermal Radiation

An organism on the Earth receives infrared thermal radiation from its surroundings. All objects at a temperature above absolute zero radiate energy. For relatively cold objects, such as biological organisms on the surface of the Earth, this radiation is comprised entirely of infrared wave lengths. These objects radiate according to the fourth power of their absolute temperature. The efficiency with which they radiate is determined by their surface characteristics as defined by an emissivity constant, ϵ. The radiation law then takes the form

$$R = \epsilon \sigma T^4 \qquad (2)$$

where T is the absolute temperature and σ is the Stefan-Boltzmann

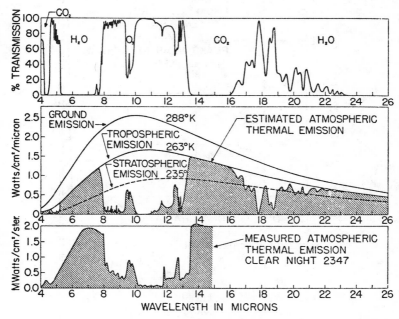

FIG. 5. Infrared absorption and emission by the atmospheric gases H_2O, CO_2, and O_3, and the thermal emission from the ground is shown as a function of the wavelength. Figure from Gates (1962).

constant. Most organisms have an emissivity between 0.95 and 1.00 and therefore may be regarded as "blackbodies." The surface of the ground radiates very nearly as a blackbody. The atmosphere radiates infrared radiation toward the ground and also outward toward space. In order to emit infrared radiation, a gas must be capable of absorbing at infrared wave lengths since a good absorber is a good emitter at the same wave length. Water vapor, carbon dioxide, and ozone have strong absorption bands at infrared wave lengths as shown in Figure 5. These absorption bands then provide efficient regions of emission for radiating streams of energy groundward and spaceward. The observed downward streams of infrared radiation from the atmosphere are shown in Figure 5. If the Earth's atmosphere was composed only of oxygen and nitrogen, without carbon dioxide (0.03% by volume) and water vapor (0.1 to 1%), the sunlit side of the Earth would be intensely hot and the nighttime side would be very cold; the climate of this planet would be most inhospitable. These truly minor constituents, water vapor and carbon dioxide, play a critical role in screening the surface from the solar infrared radiation,

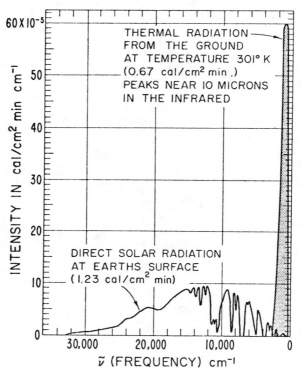

FIG. 6. The monochromatic intensity of solar radiation at the Earth's surface and the infrared thermal radiation from the ground as function of the frequency of the radiation.

blanketing the surface from the cosmic cold of space, and radiating warmth groundward. The plants and animals on the surface of the Earth are sandwiched between ground and sky receiving this flow of energy as heat and moderating their temperatures. Clear, dry nights are cold nights, and humid or overcast nights tend to be warmer because of additional infrared radiation from the sky. Animals are clearly responsive to this radiation and behave according to conditions. An important factor of past climates is the degree of cloudiness, both in the reduction of solar radiation reaching the surface and in the enhanced nocturnal radiation received within this "sandwich" between sky and earth. The importance of the long-wave infrared radiation which is ever present in the biosphere, day and night, winter and summer, must not be ignored in the consideration of biological climates. This will be part of the story to be unfolded here; the radiation climates of the Earth.

The thermal radiation from the ground which is radiated to objects immediately above it is shown in correct proportion when compared with the solar radiation on a frequency plot such as Figure 6. The frequency span of the infrared thermal radiation is not great, but the radiancy of this radiation from a surface in close proximity is substantial; namely six times that of the sun since the distance of the Earth from the sun is very large. Perhaps Figure 6 will help place these radiations in the correct perspective since one may not have thought of them in this way before.

Radiation Environments

The radiation heat load of an environment is of importance to all biological organisms within the environment and represents the flow of energy of which a portion will be absorbed by any particular organism. A few examples will now be given of these radiation environments as they occur over the globe. The radiation climate of each locality will be expressed in terms of the total radiation absorbed by the upward-facing and the downward-facing surfaces of a horizontal flat blackened plate. The radiation load on the plate will be equal to the sum of the downward stream of radiation and the upward stream of radiation. This radiation load will be the potential radiation load for any actual surface, such as a horizontal leaf or an animal located within this radiation environment. The effective radiation load will be the amount of the potential radiation load actually absorbed by the organism. This will depend upon the spectral absorptivity of the organism and will be discussed in more detail later.

The upward stream of radiation will involve the ever-present infrared thermal radiation emitted from the surface of the ground and the direct sunlight and reflected skylight which are reflected from the surface of the ground during the daytime. The downward stream of radiation will be comprised of the direct solar radiation and skylight and the omnipresent infrared thermal emission from the atmosphere. The only complete measurements of this nature, in which the contributions to the downward and upward streams of radiation were measured individually, are reported by Fleischer (1958) for Hamburg, Germany. Therefore, the first example to be described will be this one. Although there are numerous measurements available from meteorological stations throughout the world, the measurements are designed for meteorological purposes and are not of the right form nor complete enough for application to ecology. The meteorologist is interested in the net flux at the surface, which involves the difference between the downward and upward streams of radiation, while the biologist needs the sums and actually needs to know the strength of each contributing component.

Figure 7 shows the results of observations made by Fleischer throughout a 24-hour period at Hamburg, Germany, on June 5, 1954. The data are representative of the measured fluxes of radiation over an open field

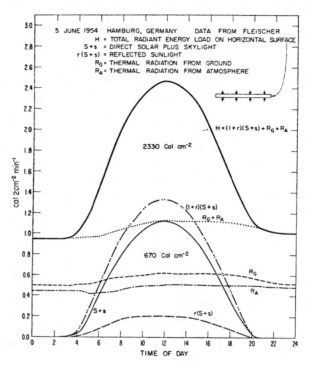

FIG. 7. The daily variation of the potential radiation heat load at Hamburg, Germany on 5 June 1954 on a horizontal surface receiving the downward stream of sunlight, skylight, and atmospheric thermal radiation and receiving the upward stream of reflected sunlight, reflected skylight, and thermal radiation from the ground. The individual fluxes are shown and the total short-wave solar radiation fluxes and the total long-wave thermal radiation fluxes are summed separately. Data obtained by Fleischer (1958).

of grass on a clear day. The strong diurnal behavior of the direct solar and skylight fluxes and the nearly constant behavior of the infrared thermal fluxes is evident from the curves. The total potential heat load on a horizontal plate also undergoes a substantial variation during the daylight hours, reaching a peak of 2.47 cal/2 cm^2/min during midday. At night, the potential heat load did not drop below 0.95 cal/2 cm^2/min. For comparison purposes, one should note that the value of the solar constant is 2.0 cal/cm^2/min.

For most localities only the direct solar radiation and scattered skylight falling on an upward facing horizontal surface have been measured with an Eppley pyrheliometer or other similar instrument. None of the other radiation components, either short or long wave, has been measured. The reflected solar and skylight from the ground were taken to be 18% of the downward stream of solar and skylight radiation for all of the

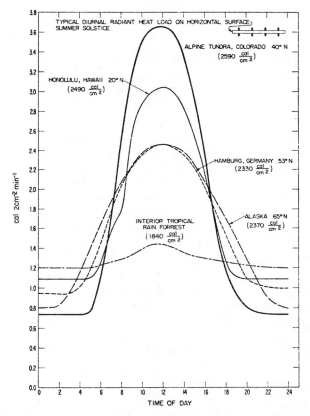

FIG. 8. Potential radiation load on a horizontal surface (upper and lower) for various localities as a function of the time of day. The total daily radiation load is also given. Only the curve for Hamburg, Germany, is based on complete observations. The other curves are based on observation of the sunlight and skylight fluxes and reasonable estimates on the long wave thermal radiation and of the reflected sunlight and skylight.

radiation climates to be described here. This is tantamount to having the same ground cover at all sites. Most meteorological stations report air temperature near the surface and the relative humidity. Using this information it is possible to obtain reasonable values of the long-wave downward flux of radiation from the atmosphere by means of the following empirical formula established by Brunt:

$$R_A = \sigma T_A^4 (0.44 + 0.08 \sqrt{e}) \tag{3}$$

where T_A is the temperature of the air near the surface and e is the vapor pressure in millibars. By assuming a diurnal change for the surface temperature of the ground, which is compatible with reported observations of ground surface temperatures *versus* air temperature, the infrared thermal radiation emitted by the ground surface was calculated according to

$$R_G = \sigma T_G^4 \tag{4}$$

where T_G is the ground surface temperature. Strictly the ground surface temperature must be used, since the ground radiates from the surface particles and not from a depth of as little as a few millimeters. Geiger (1957) and Brooks (1959) give examples of the daily variation of ground temperature *versus* air temperature.

The radiation climate for each locality was calculated using characteristic meteorological data for that locality for the summer solstice. Localities chosen were Honolulu, Hawaii, at 20°N, the interior of Alaska at about 75°N, and the alpine tundra in Colorado at 40°N. A comparison of these radiation climates in terms of the potential heat load on a horizontal flat plate is given in Figure 8. Although the "dawn comes up like thunder" in the tropics, the heat load in the alpine tundra advances like an avalanche as the sun bursts upon the mountain slope. It is immediately apparent that the alpine tundra possesses one of the hottest climates on Earth for clear or nearly clear summer days. The potential heat load on the summer solstice in the alpine at 40°N is 2370 cal/2 cm²/day, in Hamburg, Germany, at 51°N is 2330 cal/2 cm²/day (for June 5, 1954), and in Honolulu, Hawaii, at 20°N is 2490 cal/2 cm²/day. It is interesting to note that these relatively warm summer climatic conditions in different parts of the world in full sunshine have a total potential radiant heat load within about 10% of one another. In addition, the potential radiation load on a horizontal plate in the interior of a tropical rain forest is estimated at 1840 cal/2 cm²/day and will not vary substantially from season to season. However, it is also to be noted that the maximum radiation load in the alpine tundra is 3.66 cal/2 cm²/min compared with 2.46 cal/2 cm²/min at Hamburg, Germany, and Alaska. As far as the plants and animals are concerned, it is the maximum radiation load which they must withstand and so these values may have more meaning in terms of adaptation and limitation than the total integrated daily values. The daily totals may have a different significance.

One might now ask how it is that a plant is coupled to this radiation environment? It is through the absorptivity factor which depends upon the wave length of the radiation. Although it would be most desirable always to have the complete wave length or frequency distribution of the total radiation load, this is not always practical. In its stead, one can treat the solar and daylight radiation of the heat load separately from the infrared thermal radiation components. This is the reason for giving in Figure 7 the sum $(1 + r)(S + s)$ and also $R_A + R_G$. If, for example, a plant has an average absorptivity of 0.65 to sunlight and 0.95 to thermal radiation, then, in the alpine tundra, it will be coupled to the radiation load with an absorptivity of 0.95 at night and 0.75 at noon. In other words, the plant will absorb 95% of the night time radiation load and only 75% of the daytime total radiation load at noon. During the rest of the daytime the over-all absorptivity will fall between 75% and 95%. The point to be made here is that the plant will be coupled to the potential radiation to a different degree depending upon the spectral distribution of the radiation comprising the potential radiation load. Needless to say, an animal may be coupled to the same potential radiation load to a very different degree than a plant.

Reradiation and Convection

In general, there are three mechanisms operative in dissipating the radiation load on a plant or animal in still air. These are: reradiation, convection, and transpiration or evaporation. The reradiation is simply $\epsilon \sigma T^4$ where ϵ is the emissivity of the plant surface at infrared frequencies, a value between 0.95 and 1.00, and T is the leaf temperature in °K. Usually between 70 and 90% of the radiation load on the leaf will be dissipated through reradiation. In the daytime, convection will usually cool a plant or animal; at night it may add heat. Since the subject of convection is enormously complex, only a very specific aspect, namely, natural or free convection for a plant, will be discussed here. In still air any object warmer than the air will cause the air near the object to be warmed. Because air which is warmer than the surrounding air is also of lower density, this warmer air will rise, carrying heat away from the warm object and simultaneously entraining the cooler air around the object. The physicist and the heat transfer engineer have studied in great detail the phenomenon of free convection for many warm objects of simple geometry in air. The results of these studies are well known and can be found in any text book concerned with heat transfer. Since most biological organisms can be approximated by a flat plate or cylinder, Gates (1962) has summarized heat transfer theory for these geometrical configurations. Since the leaves of a broad-leafed plant can be approximated by a flat plate, it is appropriate here to write the following equation for the heat lost by convection for a horizontal flat plate in air in cal/cm²/min.

$$Q_C = 6.0 \times 10^{-3} \frac{\Delta T^{1.25}}{L^{0.25}} = h_C \Delta T \tag{5}$$

where h_c is the convection coefficient, ΔT is the difference in temperature in °C. between the plate or leaf and the air, and L is the characteristic dimension of the leaf in cm, roughly the average of the length and the width. This equation averages the situation for a warm upward-facing and a warm downward-facing surface. If the plate or leaf is 10°C. above air temperature, and if it has a characteristic dimension of 1 cm, free convection will carry away 0.08 cal/cm²/min.

In order to visualize the convection flow around the plant leaves, Gates and Benedict (1963) placed plant leaves in a schlieren optical system and

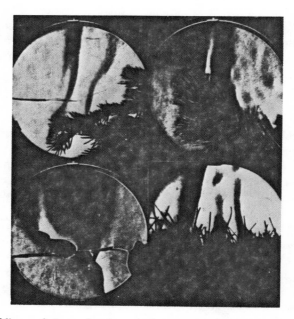

FIG. 9. Schlieren photographs of convection patterns in still air from blue spruce showing thermocouple, white pine (upper right), American elm (lower left) and grass (lower right). Each plant was irradiated with 3600 foot candles of light.

photographed the flow as shown in Figure 9 for several different leaves. These leaves were being irradiated with light and heat from a sun lamp to simulate natural conditions in full sunlight. The amount of radiation incident upon the leaf, upper and lower surface, was measured with a radiometer. The energy transported away from the leaf in the warm convection plume was measured by observing with a fine thermocouple the temperature of the plume and the surrounding air and by a determination of the volume rate of flow of the warm air in the plume. The technique is described in detail in the paper by Gates and Benedict (1963). These observations on broad leaves completely confirmed the flat plate theory ap-

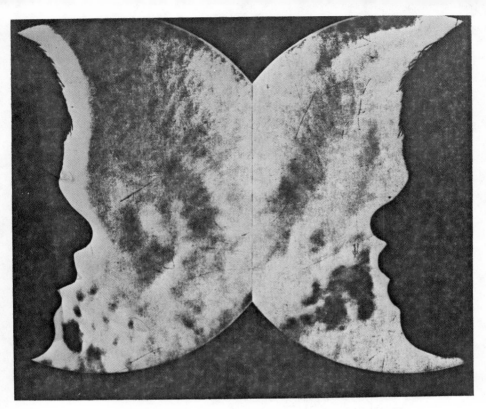

Fig. 10. Schlieren photographs of two small girls in profile showing the boundary layer of warm stagnant air near the forehead surface and the warm exhaled air from breathing and speaking.

proximation and gave original data which could not be theoretically derived for complex structures such as conifer branches.

The same technique can be applied to animals for observing the flow of air near the animal and the exchange of air and moisture during breathing. The flow of air around a bird in a wind tunnel would be particularly interesting. The most ready, able, and willing subjects, which the author had available, were his two small daughters whose profiles are shown in Figure 10. The left hand profile shows best the stagnant boundary layer around the head, forehead, and face. The warm air exhaled during breathing and speaking can be seen in front of each child.

The boundary layer of stagnant air surrounds all of us in still air and cushions us from the harshness of our immediate environment. This was realized emphatically by the author as he sat in a Finnish sauna at 230°F and came to the conclusion that he was not boiling away. Blowing on one's hand however produced a burn, for this destroyed the boundary layer and entrained the very hot air to the surface.

Radiation Load and Plant Temperature

It is now evident that an organism is coupled to the environmental air temperature by means of the convection coefficient which in turn depends upon the dimension of the organism and upon the difference between the organism's surface temperature and the air temperature. In order to combine the energy-loss mechanisms of reradiation and convection it is possible to express both of these in terms of the difference in temperature between the leaf and the air. Graphical representation of

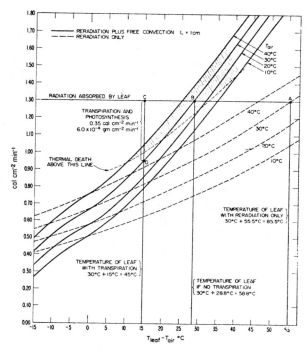

FIG. 11. Energy exchange diagram for a leaf showing the mechanisms by which absorbed radiation is dissipated by reradiation, convection, and transpiration with reference to the air temperature. If the air temperature is 30°C then the leaf temperature will be above the air temperature according to point A, reradiation only, point B, reradiation and convection only, and point D, reradiation, convection, and transpiration. The convection is computed for a leaf 1 cm wide.

the energy loss is shown in Figure 11 where the convection loss plus reradiation loss is given by the heavy lines and reradiation alone is given by the lower lighter lines. Each line as shown represents a given air temperature from 10°C. (50°F.) to 40°C. (97°F.). The ordinate is energy in cal/cm²/min and represents the effective radiation load on one surface of a horizontal leaf or the energy loss through convection and reradiation from one surface of a horizontal leaf of characteristic dimension 1 cm. If the effective radiation load for a given leaf in a specific locality is 1.30 cal/cm²/min, and the air temperature is 30°C., then, if the temperature of the leaf is controlled only by reradiation with no convection and no transpiration, the leaf will assume a temperature 55.5 °C. above air temperature or a leaf temperature of 85.5 °C. Clearly this is much too high a temperature for a leaf and the leaf must be cooled by other mechanisms in addition to reradiation. If free convection and reradiation are operative, then, with no transpiration, the leaf will assume a temperature of 58.8 °C.

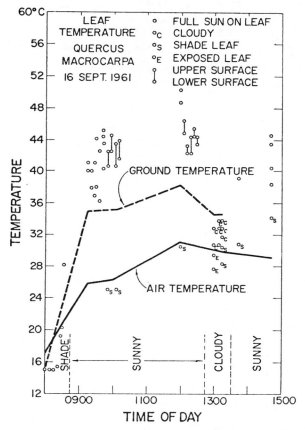

FIG. 12. Temperatures of *Quercus macrocarpa* leaves on 16 September 1961 in Boulder, Colorado as a function of the time of day. Measurements from data by Gates (1963).

which is 28.8 °C. above air temperature. This is also a rather high temperature for a leaf to withstand. In fact, denaturation of proteins and thermal death will probably occur. There can be no doubt but that transpiration plays a crucial role in keeping a plant below the lethal temperature point.

Many measurements of leaf temperatures have been made by Gates (1963) and by others who have shown that leaves in full sunlight normally have temperatures about 10°C. to 15°C. above air temperature and very seldom more than 20°C. above air temperature. Temperatures of *Quercus macrocarpa* leaves in full sunlight are shown in Figure 12. Therefore, if one observes the leaf temperature to be 15°C. above the air temperature when the air temperature is 30°C., then, in our diagram of Figure 11, this fixes precisely the position on the 30°C. air temperature line. The difference between the energy load at point C, 1.20 cal/cm²/min, and the energy load at point D, 0.95 cal/cm²/min, must represent energy taken up by transpiration and photosynthesis. A small portion of this will be taken up by photosynthesis and most of this energy dissipation will be due to transpiration which would amount to nearly 6.0×10^{-4} gm/cm²/min. If stomatal control changes the diffusion resistance to transpiration, then, as long as the air temperature remains at 30°C., the point D will slide along the 30°C. cooling curve line. Hence the length of the line CD is controlled by transpiration, which ties the entire response to the morphology of the leaf and to the availability of water to the plant. Any mechanism which can control the transpiration rate will change the length of the line CD and move the point D to the right or to the left. The transpiration rate will depend upon the stomatal control, availability of water through the root system, and the vapor pressure gradient between the leaf and the surrounding air. Hence, the vapor pressure of the air also enters into the final adjustment of the line CD. Let it suffice to say here that these various factors controlling transpiration can also be dealt with analytically, including the morphological structure of the leaf. If D should move too far up to the cooling curve line, it will cross into the zone of thermal death. In Figure 11 the temperature for thermal death is taken arbitrarily as 50°C. For some plants this is too low and 55°C. or 60°C. would be more reasonable. For other plants 50°C. is too high.

Using an energy exchange diagram, such as Figure 11, it is possible to trace out the entire diurnal behavior of a leaf in response to a varying radiation load, a varying air temperature, a changing relative humidity, and changing stomatal control. A hypothetical diurnal cycle is traced out in Figure 13. Just after dawn, about 0600, the effective radiation load on one surface of the leaf is taken to be 0.64 cal/cm²/min and the air temperature about 18°C. The leaf is transpiring slightly and adjusts to a temperature 1.2°C. above air temperature or a leaf temperature of 19.2°C. By 0800 the effective radiation load has become 0.82 cal/cm²/min, the transpiration rate has increased substantially, and the leaf temperature is now 7.5°C. above the air temperature, which is up to 23°C., or a leaf temperature of 30.5°C. Throughout the morning the effective radiation load, air temperature, transpiration rate, and leaf temperature increase until at noon, 1200, the leaf temperature is 10°C. above the air temperature, which is now 30°C., giving a leaf temperature of 40°C. In the early afternoon the effective radiation load may begin to diminish, but the air temperature may still be increasing. If, for example, stomates close down then the leaf temperature will increase substantially as shown by the point for 1400 in Figure 13 at which time the leaf temperature will be about 43.3°C. During the afternoon the radiation load continues to drop, the air temperature, the transpiration, and the leaf temperature also drop, with the points falling consistently to the left in Figure 13.

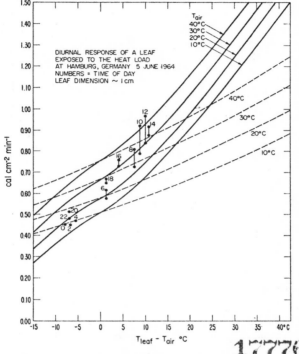

FIG. 13. Energy exchange diagram for a leaf of width 1 cm showing the diurnal cycle of effective radiation load, the transpiration during daylight hours, and the leaf temperature. The points are labeled according to the hour of the day.

Then, during the night, a somewhat different situation prevails. Radiation loss from the leaf to the cold, clear night sky would tend to cool the leaf substantially below air temperature if convection did not occur. But now, contrary to the daytime situation, the air being warmer than the leaf contributes heat to the leaf by convection. The points fall directly on the cooling curve for a given air temperature. The cooling curve now is to the right of the reradiation curve rather than to the left as for the daytime conditions in full sunlight. Hence, the nighttime points at 2000, 2200, 0000, and 0200 are shown with dropping air temperature. Then at 0400 with a slightly warmer air temperature the sun then breaks over the horizon and the daily cycle repeats in various forms, depending upon whether the day is clear or cloudy.

Although a leaf was used to illustrate the response of an organism to a radiation load, the same procedure could be used for an animal. One would need to take account of the mobility of the animal as it moved from one environment into another. It is also possible to treat forced convection in the same manner and to formulate a "cooling" curve which takes account of the influence by the wind.

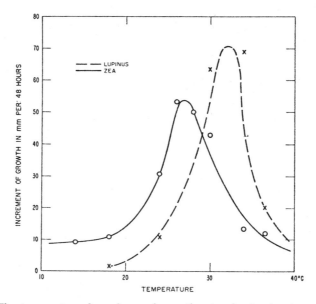

FIG. 14. The temperature dependence of growth rate of a temperate region plant, *Lupinus*, and of a warm region plant, *Zea*. From Thompson (1942).

Leaf temperature has considerable significance in terms of chemical rate processes going on within the leaf which are temperature dependent. It is well known that organisms have an optimum temperature for certain biological activity and that conditions are less favorable toward lower and higher temperatures. Evidence for such an optimum in plants is shown in Figure 14. Many other examples are well known for other organisms and these are thoroughly discussed by Johnson, Eyring, and Polissar (1954). These physiological processes are controlled by enzymes and other proteins in a complex manner; nevertheless, certain definite reaction rates are proceeding and are producing an end result which is temperature dependent.

Chemical processes, in which certain end products are formed from initial products, depend upon a quantity called activation energy which represents a potential barrier between the initial and final products. Also involved is a description of the number of activated states of a molecule or system at a particular temperature in terms of the so-called partition function. Together, these concepts describe the ability of a chemical reaction to proceed forward or backward and to reach a definite equilibrium. Since many biological processes are chemical, it is important that the theory of rate processes be kept in mind when discussing biological response. The fact that many biological responses have an optimum rate at a certain temperature would indicate that there is a favorable temperature regime at temperatures less than the optimum temperature and an unfavorable regime at temperatures above this. Throughout the favorable temperature regime, certain chemical processes are forming favorable end products, such as proteins, for the organism and throughout the unfavorable regime these products are being destroyed in processes such as the denaturation of the proteins. At very low temperatures, all reactions may cease and at too high a temperature total destruction of organic complexes will occur and death will ensue.

The energy flow interpretation of climates and environment permits a direct coupling to reaction rate theory, for now one can predict the temperature an organism will assume in a given environment and, from temperature, one can discuss the response of the organism in terms of the kinetic basis of molecular biology. Problems of growth and productivity, life-span, genetic mutation and evolution, physical activity and adaptation, and many others can be understood through detailed knowledge concerning reaction rates, temperature, and energy flow. These biological problems are complex and must be treated with all the sophistication of modern science. Ecology, the interaction of organisms with their environment must be understood from the standpoint of energy flow, organism temperature, diffusion theory, chemical rate processes, and modern molecular biology.

REFERENCES

Brooks, F. A. 1959. *An Introduction to Physical Microclimatology*, Univ. Calif., Davis, Calif. 264 pp.
Fleischer, R. 1958. Die atmosphärische Gegenstrahlung. *Medizin-Meteorologische Hefte, 13*, 142–148.
Gates, D. M. 1962. *Energy Exchange in the Biosphere*. Harper and Row, Inc. New York.
——— 1963. Leaf Temperature and Energy Exchange. *Archiv Meteor. Geophys. u. Bioklim.*, B, *12*, No. 2, 321-336.
——— and C. M. Benedict, 1963. Convection from plants in still air. *Amer. J. Bot.*, *50*, July issue (in press).
Geiger, R. 1957. *The Climate Near the Ground*, Harvard Univ. Press. Cambridge, Mass.
Johnson, F. H., H. Eyring, and M. J. Polissar, 1954. *The Kinetic Basis of Molecular biology*. John Wiley and Sons, Inc. New York. 874 pp.
Thompson, D'Arcy, 1942. *On Growth and Form*. Cambridge Univ. Press.

ENERGY DYNAMICS OF A FOOD CHAIN
OF AN OLD-FIELD COMMUNITY

FRANK B. GOLLEY

INTRODUCTION

In recent years there has been a growing interest in the study of the transfer of energy through natural systems (ecosystems, Tansley 1935). Park (1946) stated that "probably the most important ultimate objective of ecology is an understanding of community structure and function from the viewpoint of its metabolism and energy relationships." Aquatic biologists have taken the initiative in the study of community energetics, and most of the information available today concerns fresh water or marine communities. A great need exists for similar studies on terrestrial communities.

In this study a food chain of the old field community, from perennial grasses and herbs to the meadow mouse, *Microtus pennsylvanicus pennsylvanicus* Ord, and to the least weasel, *Mustela rixosa alleghenieusis* Rhoads, was chosen for investigation. This food chain included the dominant vertebrate of the community (*Microtus*) and one of its main predators (*Mustela*) but excluded the otherwise important insects, other invertebrates, bacteria, and fungi. The primary objectives of the study were to determine (1) the rate of synthesis of organic matter by the primary producers—the vegetation, (2) the path of this energy from the vegetation through the mouse to the weasel, and (3) the losses of energy at each step in the food chain.

The writer wishes to acknowledge with gratitude the suggestions and guidance of Dr. Don W. Hayne, Institute of Fisheries Research, Michigan Department of Conservation, especially concerning that portion of the study dealing with the population dynamics and productivity of the *Microtus* population. The writer also thanks Dr. John E. Cantlon, Department of

Botany, and Dr. Robert C. Ball, Department of Fisheries and Wildlife, Michigan State University, for aid given throughout the project. The investigation was supported by the Michigan Agricultural Experiment Station through a project administered by Dr. Hayne.

DESCRIPTION OF THE AREA

The study area was located in a large field on the Michigan State University State Farm approximately one mile south-east of Okemos, Ingham County, Michigan (sec. 27, T. 4N, R. 1W). As far as is known, this farm was last tilled in 1918 when it was given to the State of Michigan by Mr. John Fink. It was acquired by Michigan State University in 1940 and was pastured from 1940 to 1942. The study area has been undisturbed since 1942, with the exception of some tree planting by the Department of Forestry, Michigan State University, and probably occasional burning. The tree plantings appeared to be only slightly successful. The vegetation on the area was unburned from January 1952 to March 1957.

The field in which the study area was located was situated on the north terrace of the Red Cedar River, approximately 20 ft above the level of the river. The topography was gently undulating, with a relief of 15 ft or less. A shallow depression ran through the center of one of the trapping areas and served as a drain during the heavy rains in the winter and spring. On February 9, 1957, the snow melt-water was approximately 7 in. deep in this drainage area. As the snow melted in February and March much of the study area was inundated, with grass hummocks and hillocks on the border of the trap area providing the only dry sites.

The soils on the study area were predominantly Conover and Miami loam (determined from the soil map by Veatch *et al.* 1941).

On the east the field was separated from similar habitat by a paved county road. The north boundary was predominantly pasture land and orchard. The west boundary was an experimental alfalfa field left uncut in 1957. To the south the field was bounded by an unused gravel road, which ran along the ridge top above the river terrace and separated the field from other old field vegetation containing more woody cover and indicating a later stage of old field development. The field itself contained approximately 10 ha of relatively homogeneous habitat.

The climate in this area of Michigan is characterized by cold winters and mild summers (Baten & Eichmeier 1951). Yearly precipitation at East Lansing (1911-1949) averages 31 in.; growing season (last day in spring to the first day in fall when the temperature reaches 32°F) precipitation averages 17 in. The mean annual temperature is approximately 47°F, with extremes ranging from −20° to +102°F. The growing season averages 147 days. Solar radiation at East Lansing (3 mi west of the study area) is peculiar in that a plateau in the insolation curve may be expected about April 25 to May 20. When solar energy received at East Lansing is compared with that at most of the 92 weather stations in North America measuring solar insolation, it is evident that East Lansing receives annually less solar heat than any other station, with the exception of Fairbanks, Alaska (Crabb 1950b).

No attempt was made to make a complete survey of the flora and fauna of the community. The vegetation of the study area was transitional between the perennial grass stage (perennial grasses predominant) and the perennial herb stage (perennial herbs co-dominant with the grasses) of old field succession (Beckwith 1954). The vegetation is considered similar to the bluegrass-upland association of Blair (1948) and the upland community of Evans & Cain (1952).

Canada blue grass, *Poa compressa*[1], was dominant over the entire area, with three herb species, *Daucus carota*, *Cirsium arvense*, and *Linaria vulgaris*, sharing dominance in portions of the area. The study area was divided into four facies on the basis of co-dominance of the above herbs with *Poa compressa*. Mosses, undeveloped small herbs, and grass shoots formed a subordinate layer beneath the grass and perennial herb layer. A woody overstory occurred sporadically over the area, consisting primarily of *Crataegus* spp., *Pyrus communis*, *Prunus pennsylvanica*. The woody plants were a relatively unimportant component of the vegetation, the percentage cover for all woody plants averaging approximately 0.5.

The vertebrate dominants of the community, excluding birds, were *Microtus pennsylvanicus* and *Blarina brevicauda*, when total number observed was used as the criterion of dominance.

METHODS

As energy flows through a terrestrial food chain there is a successive transfer and loss of energy at each step in the chain (Fig. 1). As a result of the

[1] Authorities for vascular plant binomials are those given in Fernald (1950).

continual loss of energy through respiration and through nonutilization of food, each successive population is faced with a smaller energy source. In this report the writer's approach has been to study the energy flow through each separate population, rather than to emphasize the energy exchange through the food chain as a whole. In the traditional style of presenting research methods and results separately, this concern for each species population becomes especially evident and necessarily obscures the picture of energy flow through the entire food chain. The writer believes that this method of presentation is most satisfactory for an exploratory study of this nature. However, by referring to Fig. 1, the reader will be able to follow the flow of energy through the food chain without difficulty.

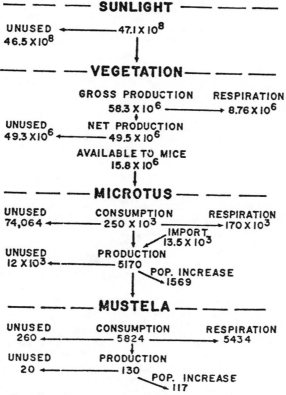

FIG. 1. The energy flow through the food chain from May, 1956 to May, 1957, on one hectare. All figures are Calories per hectare. Solar input represents that for the 1956 growing season.

Measurement of Solar Insolation

Records of solar insolation (the rate at which solar energy is received on a horizontal surface at the surface of the earth) were obtained from the Michigan Hydrologic Research Station (the Agricultural Research Service, USDA, and the Michigan Agricultural Experiment Station cooperating). The station operates an Eppley ten-junction thermopile, thermoelectric pyrheliometer, mounted on a small instrument house on an isolated section of the Michigan State University Farm, East Lansing. There is little smoke contamination at this location (Crabb 1950a) and it is assumed that the records obtained at the pyrheliometer were applicable to the study area approximately 3 mi to the east.

Studies of Vegetation

The vegetation was studied from August 1956 to September 1957. Midway through the investigation (March 1957) a fire destroyed the vegetation on one-half of the study area, and it was necessary to move the entire operation to another portion of the same field. Fortunately there was no discernible difference in vegetation or topography at these two locations.

Square clip-plots (0.5 × 0.5 m) were used to estimate the standing crop of vegetation. All plant stems within the quadrats were clipped at the ground level. The cut vegetation was transferred to plastic bags and transported to the laboratory, where live grasses, live herbs, and dead vegetative materials were separated according to species. These materials were then dried at 100°C for 24 hrs and weighed. Monthly data were averaged and the standing crop of vegetation was expressed in grams of dry weight per 0.25 m^2. Ten to twenty random clip-plots were chosen for investigation each collection period, with the exception of late March. The adverse weather conditions in the latter part of March allowed an estimate to be made of green vegetation on only two plots.

Samples of several months' collections of dried grasses and herbs were randomly chosen for calorific analyses. These materials were ground in a Wiley Mill. Three subsamples from each species sample were analyzed in a Parr adiabatic bomb calorimeter.

The standing crop of roots and above-ground portions of the plants which escaped clipping were also determined at three periods during the 1957 growing season. At each collection, five of the plots which had been clipped were chosen at random and a 225-cm^2 piece of sod was cut from each plot. The sod was later washed in running water, oven dried at 100°C and weighed to determine average weight of roots per square meter. The samples were taken to a depth of 15 cm and did not represent the complete root biomass. Although Shively & Weaver (1939) show that the roots of many prairie plants extend at least several feet into the soil, for this study it was assumed that the main mass of roots in the old field community were concentrated within 15 cm of the surface.

The standing crop of vegetation measured here was less than the total amount of organic matter synthesized over the growing season. This is because the standing crop includes neither the amount of vegetation which grew and died between the periods of measurement, nor the amount of vegetation consumed by animals. In this study, some dead vegetation of the current growing season was unavoidably included with the green vegetation since no effort was made to separate the dead and living portions of one leaf or small plant. The material that grew and died back to the ground during the current growing season could not be separated from the dead vegetation and is therefore a source of error in determination of net production. The magnitude of this error was not estimated.

The amount of vegetation consumed by *Microtus* was estimated from feeding experiments and stomach sample analyses. The food consumption of herbivorous insects and other invertebrates, which in the pasture community may be of considerable magnitude (Wolcott 1937), was undetermined.

Another source of error in estimating the net production of the vegetation is pointed out by Pearsall & Gorham (1956). They suggest that in perennial vegetation the peak standing crop is formed from (1) the accumulation of the organic matter during the present season and from (2) the organic matter stored in the roots the previous season. The techniques used in this study allowed no estimate of the contribution of the previous season's production to the current peak standing crop.

To obtain a complete estimate of the energy utilization of vegetation (gross production, Odum 1956), the energy used in respiration of the vegetation must be added to the net production. A field respirometer was devised to make a rough measure of the respiration of the vegetation and the soil organisms. Two 5-qt oil cans were forced 5 cm into the ground at randomly chosen sites on the study area. Gases were withdrawn from the cans into Bailey gas analysis bottles. Carbon dioxide and oxygen content of the air in the cans was determined in an Orsat-Henderson gas analysis apparatus. The cans were placed in the ground and the first samples were withdrawn approximately 1 hr after dark. The second sample was taken in the early morning when the air temperature at the ground surface was approximately equal to the air temperature observed when the first samples were withdrawn the previous night. The consumption of oxygen and production of carbon dioxide over the night (12 hrs) was determined as the differences in the percentage composition of oxygen and carbon dioxide in the air in the cans at the first and second sampling. No measurement was made of the diffusion of gases between the soil and the air under the cans. The RQ (CO_2/O_2) was calculated and the thermal equivalent (in Calories) of the oxygen used and the carbon dioxide produced was extrapolated from the tables in Brody (1945). Respiration was then expressed as Calories used per gram of plant tissue per hour respiration during the night.

STUDIES OF THE *Microtus* POPULATION

The energy dynamics of the *Microtus* population were studied both in the field and by laboratory experiments. In the field a live-trapping program was initiated in May 1956 and continued to September 1957. This paper includes only the data for one annual cycle, May 1956 to May 1957. The program was designed to yield information on population density, mortality, growth rate, and production of young. The trapping design was developed by D. W. Hayne and consisted of two crossed trap lines, 100 m long, with the live traps spaced 2 m apart. One trap line was operated for 24 hrs, it was then unset and the line crossing it was set for 24 hrs. The total trapping period extended for 6 days, with 3 days for each line. Since by the sixth day of trapping unmarked animals were generally caught only in the end traps in a line, it was assumed that six days of trapping was adequate to capture most of the animals living in the trap area. It was desirable to use as short a trapping period as possible because trap mortality tended to increase progressively during the trapping period.

Captured animals were toe clipped, sexed, aged, weighed, and examined for breeding condition before being released. Traps were baited with oatmeal, and during the colder months corn was also placed in the traps to serve as a high energy supplement. It was thought that the use of corn materially reduced trap mortality during the winter. Covers, made of asphalt shingles covered with aluminum foil, shielded

TABLE 1. Dynamics of the *Microtus* population on one hectare

Date	Average Number Captured Per Line	Fraction[1] Per cent	Population Density Per Hectare Numbers	Size of Trap Area Hectares	Average Individual Weight Grams	Standing Crop per Hectare Calories[2]	Mortality Between Successive Trapping Periods	
							Rate[3]	Calories
May 22-28	4	29.4	5.2	2.6	29	205	75	393
July 24-30	12	41.7	9.6	3.0	29	382	37	232
Sept. 4-10	20	41.0	16.5	3.0	29	654	52	539
Oct. 9-15	19	32.4	21.2	2.7	31	898	42	710
Nov. 17-23	28	22.2	42.4	2.4	29	1685	39	1162
Jan. 6-23	22	(13.3)[4]	77.3	2.1	27	2847	27	1245
Feb. 21-27	39	13.0	139.1	2.1	29	5525	(55)[4]	3284
March 19-26	22	13.6	75.3	2.2	25	2578	35	937
April 25-1	43	30.4	53.2	2.6	32	2330	57	1931

[1] The percentage of total captures taken in common to both lines.
[2] Caloric value of *Microtus* was 1.37 Calories per gram.
[3] Percentage disappearing from one trap period to the next.
[4] Estimated values.

the traps from sunlight, rain, and snow and were thought to reduce trap mortality especially in the summer.

A ratio method was used to estimate population density. The following formula was suggested by D. W. Hayne:

$$P = \frac{bc}{ad} \quad (1)$$

where a is the number of animals captured in common to both lines, b the average number of animals captured in each of the two lines (both a and b exclude those animals dying in the traps during the trapping period), c average for the two lines of all captures including deaths due to trapping, d the effective area trapped, and P the population density per unit area.

The trap area, bisected by each trap line, was a square of 10,000 m². Since the home ranges of the mice extended an unknown distance beyond these lines the square was increased on both sides and ends. On each side the area added was set arbitrarily as the fraction $\frac{a}{b}$ (in Formula 1) times one-half the trap line length, 50 m, times the length of one side, 100 m. At each end of the trapping square, the area was increased by a semicircle with a radius of one-half the length of the trap line plus the fractional increase computed above. Information on numbers captured and population estimates are shown in Table 1.

In calculating the production of (or total weight grown by) the *Microtus* population, it was necessary to use one method for the animals which were susceptible to capture (the adults and an unknown proportion of the juveniles) but to use an entirely different method for the nestling young which do not enter traps. For the trapped animals, production was calculated from the rates of growth and mortality observed among the trapped individuals, while for the nestling young, production was inferred from the observed rate of pregnancy, the known rate of growth of young, and the calculated rate at which the young entered the trap-susceptible population.

In calculations of production, the use of instantaneous rates has been advantageous (Clarke 1946, Clarke Edmondson & Ricker 1946, Ricker 1946). Assuming constant rates, the products of the instantaneous rates and the mean population for a period will yield, respectively, the production of animal tissue by the population and the quantity of tissue lost from the population. This approach is especially useful here since it allows estimation of the biomass or number of animals which were produced, grew, and were lost between measurements of the standing crop.

The recapture of resident animals in two consecutive trapping periods allowed an estimate to be made of the rate of weight gain or loss of the individuals. This rate was estimated separately for the animals in a number of 10-gm weight classes, since the rate of growth changes with body weight (Table 2). For each weight class, the daily instantaneous rate of growth was multiplied by the mean biomass for the period to calculate the daily production by growth or the daily weight loss in that particular weight class. A proportionate weight change of less than one indicated that the animals in that class lost weight (Table 2).

To determine the production of the nestling young, it was necessary to establish (1) the potential production of young by the population (potential natality), (2) the number of young which entered the trap-susceptible population, and (3) the growth rate of the young. The estimate of the potential production of young was based on the rate of pregnancy determined in the field and on the gestation period and litter size as reported in the literature. Pregnancy rates were established by abdominal palpation of all females. Davis (1956) suggests that pregnancy is "visible" for 18 days in small rodents—this would mean that 86% of the pregnancies could be determined by abdominal palpation. Although Davis's findings could probably be extended to *Microtus*, the percentage of adult females found to be pregnant in this study (Table 3) was so large during most of the breeding season that Davis's correction could not be made. The average number of days required for a female in the population to produce one litter can be found by dividing the gestation period (21 days, Hamilton 1941) by the proportion of females pregnant, here termed f. With the additional assumptions that there is an even sex ratio and an average of five young per litter (Hamilton 1941, Hatt 1930, Blair 1940), we can infer that in each time interval of $21/f$ days, one female increases to 3.5 females—2.5 young females plus the mother. With the use of the following formula it is then possible to calculate the production of young:

$$\frac{\text{nat log } 3.5}{\frac{21}{f}} \times t \times p' \qquad (2)$$

TABLE 2. Tissue production of juveniles and adults per hectare.

Interval	PROPORTIONATE WEIGHT CHANGE WEIGHT CLASSES IN GRAMS				DAILY INSTANTANEOUS RATE OF GROWTH[1]		Growth Calories	Weight Loss Calories
	11-20	21-50	31-40	41-50	+	−		
May......	data not available							
July......	data not available							
Sept......	2.06	1.00	1:10	—	3.08	—	74	—
Oct.......	1.86[2]	1.20	1.00	.96	3.03	.19	81	5
Nov.......	1.70	1.04	.82	.69	1.32	2.28	56	95
Jan.......	—	.99	.86	—	—	5.15	—	123
Feb.......	—	—	.94	—	—	8.20	—	152
March....	1.71	1.27	.99	.91	20.92	.68	516	17
April.....	1.87	1.21	1.03	.91	9.92	1.72	190	33

[1] Positive and negative daily instantaneous rates of growth derived from proportionate weight gains and losses of all weight classes within the month.
[2] Boldface growth rates were calculated by the increment method.

TABLE 3. Potential production of new *Microtus* per hectare.

Date	Time Interval Between Trapping Days	Pregnancy Rate Per Cent	Mean Adult Population Numbers	POTENTIAL PRODUCTION	
				Numbers	Calories
May...........	64	90	7.2	24.8	102
July..........	42	90	12.6	28.5	117
Sept..........	35	90	18.5	34.8	143
Oct...........	39	90	30.7	64.5	265
Nov...........	61	65	58.2	137.9	567
Jan...........	35	00	104.7	—	—
Feb...........	27	00	103.7	—	—
March........	36	11	63.5	15.2	63
April.........	28	92	51.5	79.3	326
Totals......				385.0	1583

in mouse tissue due to the nestlings between the two trapping periods. The contribution of nestling young is shown in Table 4.

TABLE 4. Growth of nestling young per hectare.

Date	Mean Biomass Grams	Growth Rate	TOTAL GROWTH	
			Grams	Calories
May........	144	5.33	242	331
July........	135	5.33	226	309
September...	180	5.33	302	414
October.....	333	4.67	516	706
November...	893	7.67	1231	1686
March......	100	4.67	154	211
April.......	361	3.30	435	596
Totals....			3106	4253

where f is the percentage pregnancy among females, t is the time between trapping periods, and p' is the mean population.

The number of young entering the trap-susceptible population was assumed to equal the number of adults and juveniles lost to mortality plus the number needed to fulfill the population increase.

The growth rate of the nestling young was based on the weight increase from an assumed birth weight to the weight of the lightest juvenile captured in the live traps between trapping periods. Whitmoyer (1956) showed that the mean birth weight of laboratory *Microtus pennsylvanicus* was approximately 3 gm. The lightest weights of live-trapped juveniles ranged from 10 to 16 gm. The products of the instantaneous rate of growth of the nestling young and the mean biomass of young yielded the increase

When the amount of energy leaving the *Microtus* population through respiratory processes is added to the production of tissue, the result is a measure of the assimilation of the population. Assimilation is defined here as the energy which enters the population and is actually used in productive or maintenance processes. An estimate of the respiratory energy loss was made by studying the metabolic rate of *Microtus* by the McLagan-Sheahan (1950) closed circuit method. This technique utilized a series of desiccator jars connected to a pure O_2 source, a vacuum pump, and mercury manometers. Soda lime, in the bottom of jars, absorbed CO_2. The system was of known volume (approximately 2600 ml) and was kept at a constant temperature of 26°C. Wild *Microtus* were trapped the day before the experiment and fasted over night (12 hrs). Three or four mice of the same sex and weight were placed in each jar and, after air was evacuated from the jars to a negative pressure of 200 mm Hg, pure O_2 was introduced until pressure returned to equilibrium. As the O_2 was consumed in the jars, the pressure changes were measured on mercury manometers. The mice were allowed about 30 min to become accustomed to the apparatus before readings were made on the manometers. After this initial period, the mice were maintained in the jars for 1 hr. A respiratory quotient (RQ) of .85 was assumed in the computations of metabolic rate.

Metabolic rate determined by this method can not be considered a basal rate (BMR) because the animals were slightly active in the jars during the experiments. Rather than basal rate, this study determined the fasting metabolic rate (FMR) of *Microtus*. The FMR (Table 5) was calculated in terms of cc of O_2 consumed per gm mouse tissue per hour and in Cal per 24 hrs per individual mouse. The respiration of the adult biomass was estimated by multiplying the metabolic rate in Cal per 24 hrs per mouse by the population density at a trap period and by the number of days between succeeding trap periods. The product of the metabolic rate of the young (assumed to be 1.7 Cal per 24 hrs), the mean population of young, and the time interval between trapping periods yielded the respiration of the nestling young.

TABLE 5. Respiration of experimental animals.

Date	Ave. Weight Individuals Grams	Oxygen Consumption cc/gm/hr	Calories Per 24 Hours Per Mouse
Oct.	25.5	2.83	8.2
Feb.	34.9	2.55	10.6
April	31.0	2.83	10.2
Average	29.9	2.75	9.7

The FMR when applied to animals in the field should be considered a minimal figure of metabolism. Brody (1945: 477) considers that the maintenance energy expense is twice the basal metabolic rate. The energy cost of maintenance is the net dietary energy needed to carry on life processes, excluding the production of flesh, milk, or young.

To determine the caloric value of *Microtus* carcasses, four wild mice were sacrificed, minced, and dehydrated in a lyophilizing apparatus. This dried material was then burned in the bomb calorimeter to determine average caloric value per gm of dry mouse tissue. The wild mice were of average weight, ranging from 10 to 39 gm, and did not exhibit large fat deposits around the internal organs.

Food consumption by mice was studied both in the field and in the laboratory. For the field studies, wild *Microtus* were snap-trapped every three months in other areas which were characterized by a bluegrass-perennial herb vegetation similar to that found on the study area. At least 24 mice were captured during each trapping period (Table 6). These mice were brought into the laboratory and their stomachs were removed and weighed. A portion of the stomach contents was placed on a glass slide with several drops of Turtox CMC-10 mounting media. A smear was made of this mixture and, after a cover glass was placed on the slide, the slide was examined under the low power objective of a microscope.

TABLE 6. Percentage importance of food materials from stomach samples.

Food Material	Fall	Winter	Spring	Summer
Number of stomachs	35	27	31	14
Grass	54	75	74	54
Herbs	28	18	23	44
Insects	T	T	1.8	T
Fruits	17	3	.1	1
Wood	1	4	.3	—
Seeds	T	—	.6	T
Moss	T	—	.1	T
Fungi	T	—	T	1
Grasses and herbs alone				
Grass	66	81	76	55
Herbs	34	19	24	45

A stomach content key was devised by feeding in the laboratory 5 *Microtus* on diets of natural foods, each mouse receiving only one food substance. These animals were sacrificed and slides of their stomach contents served as a reference key when examining the stomachs of wild mice. Under the microscope it was possible to distinguish the following food types: grasses, herbs, woody materials, roots, seeds, fruits, mosses, fungi, and insect remains. These identifications were made on the basis of cell shape, cell wall structure, arrangement of stomata, presence of parenchyma cells, sclereids, tracheids, and other elements. An estimate was made of the percentage that each food type contributed to the bulk of the plant material on each slide. The percentage importance of the food types was determined for the collection period by averaging the data for each individual stomach.

The quantity of food consumed was measured for caged mice in two laboratory experiments. In the first experiment, 15 mice in 5 cages were fed a "standard" laboratory diet of lettuce, carrots, and oatmeal (Whitmoyer 1956) for 30 days. In the second experiment, 5 mice were maintained on fresh-cut alfalfa for 30 days. Water was available in both experiments. Animals gained weight, bred, and gave birth to normal litters on both diets. Food materials were weighed in and out of the cages daily; the weight loss of fresh food between weighings was determined by using a control cage. The information on food consumption is summarized in Table 7. The caloric value of the standard diet was determined from Wooster & Blanck (1950) and that of the alfalfa by combustion in the bomb calorimeter.

TABLE 7. Daily food consumption of individual mice on experimental diets.

	Standard Diet				Alfalfa Diet Alfalfa
	Lettuce	Carrot	Oatmeal	Total	
Consumption gms wet wgt...	24.8	10.2	4.4	39.4	28.1
Consumption gms dry wgt...	1.3	1.2	4.0	6.5	12.0
Caloric value of food per gm wet wgt..................	.18	.45	3.96	—	—
Caloric value of food per gm dry wgt..................	—	—	—	—	4.1
Calories consumed..........	4.5	4.6	17.4	26.5	49.3
Ave wgt mice (gms).........				46.0	46.0
Gms food consumed per gm mouse tissue.............				.14	.26
Food Calories consumed per gm mouse tissue.........				.58	1.07

The digestibility of the experimental diets was studied by collecting mouse feces in the cages for a 5-day period during each experiment. The feces were oven-dried and the caloric value determined in the bomb calorimeter. By this method it was possible to estimate the amount of gross energy in the feed which was undigested.

STUDIES OF THE LEAST WEASEL

The energetics of the least weasel were given a more superficial treatment than those of the vegetation, or of the *Microtus* population. Population estimates were inferred from the capture of weasels in live traps during the mouse trapping program and from counts of weasel tracks in the snow during December, January, and February. During any one trapping period, captured individuals were identified by weight under the assumption that only one individual of a particular weight would be present on the trap area. On the basis of trap records and tracking observations it appeared that there were two adult weasels on the area of 2.5 ha in the late fall of 1956. It was assumed that these weasels had been present and had produced young during the summer. The number of litters produced per year (two) and the number of young per litter (five) were accepted as reported by Burt (1948) and by Hall (1951). June and August were arbitrarily chosen as the birth dates of the litters.

Since no data on the growth of the least weasel were available, growth rates of adult and young weasels were estimated from the weights of captured animals. The initial weight for the adults in the early summer of 1956 was assumed to be 46 gm (average of 4 captures of young adults). These adults were assumed to have grown to an average adult weight of 60 gm by August (based on one capture) and to have maintained this weight through the spring of 1957. It was further assumed that the birth weight of the young was 3 gm and that each individual in the litter grew approximately 6 gm per month for the first 5 months and then 3 gm per month for the next 5 months. As with *Microtus*, production measurements were based on the instantaneous rate of growth of adults and young. Production was calculated separately for three 4-month periods (Table 8). Biomass measurements were converted to their caloric equivalent by using the factor obtained for mice carcasses.

Mortality was arbitrarily estimated as a loss of approximately 5 young, weighing 15 gm each, from September to December, 1956, and of one young, weighing 35 gm, from January to May, 1957.

Food consumption and digestibility of food were studied in the laboratory with one captured weasel (Table 9). In the course of two feeding experiments, one live mouse was placed in the weasel cage daily. The remains of the dead carcass of the mouse fed the previous day were transferred to a hardware-cloth envelope within the cage to indicate evaporation loss from the carcass. White mice (*Mus musculus*) were fed in the first study and laboratory-raised *Microtus* in the second study. Each experiment was run for a total of 30 days.

During the feeding experiment using *Microtus*, feces were collected for a 6-day period to measure food digestibility. The caloric value of these feces was determined in the bomb calorimeter.

The metabolic rate used for the least weasel was that obtained by Morrison (1957).

RESULTS

SOLAR ENERGY

The annual insolation per ha for 1956 and 1957 is shown in Table 10. Baten & Eichmeier (1951) indicate that the average agricultural growing season at East Lansing is from May 8 to October 4, with extremes ranging from April 8 to November 16. Field observations suggested that the growing season for natural vegetation of the old field was slightly longer than that for cultivated crops and extended from approximately April 1 (when spring plant growth became obvious) to approximately November 1 (when the accumulated production peak was reached in 1956). The total insolation during the 1956 growing season was 94.2×10^8 Cal per ha. Since approxi-

TABLE 8. Dynamics of least weasel population on one hectare.

Season	Average Population Number		Total Biomass Grams		Proportionate Weight Change		Production Grams		Mortality Grams		Respiration Loss Calories
	Adults	Young	Adults	Young	Adults	Young	Adults	Young	Adults	Young	
May-Aug.	.80	2.0	42.4	10.3	1.30	2.58	11.1	10.0	0.0	0.0	1091
Aug.-Dec.	.80	2.8	48.0	34.9	0.00	3.04	0.0	38.8	0.0	31.5	1884
Jan.-May	.80	1.6	48.0	61.9	0.00	1.76	0.0	35.5	0.0	10.5	2459

TABLE 9. Food consumption of the least weasel fed on live mice.

Species	Food Consumption		Caloric Value of Feces	Per Cent of Food Digested
	Grams	Calories		
Microtus	14.7	19.99	2.02	89.9
White mice	15.1	29.50	—	—

TABLE 10. Solar insolation on the study area in calories per hectare per month for 1956 and 1957.

Month	1956	1957
January	4.2×10^8	5.0×10^8
February	7.3	6.1
March	9.5	11.0
April	10.8	9.9
May	14.7	13.9
June	17.8	16.8
July	15.3	18.2
August	13.6	15.2
September	12.2	11.7
October	9.8	7.9
November	4.5	3.9
December	2.7	3.8
Total	122.4	123.4
Total growing season (April 1 to Oct. 31)	94.2	93.6
Growing season correction[1]	47.1	46.8

Data from the Michigan Hydrologic Research Station of the USDA and the Michigan Agricultural Experiment Station.
[1] 50 per cent of the total insolation during the growing season to allow for ultraviolet and infrared radiation which are not utilized in photosynthesis.

mately 50% of the incident energy (that in the ultraviolet and infrared portions of the spectrum) is not used by plants in photosynthesis (Terrien, Truffaut & Carles 1957, Daubenmire 1947), the total growing-season insolation was divided by two to give the usable insolation available to the plants. The data presented in Table 10 represent total solar insolation at the ground surface, and at the bottom of the table is shown the 50% correction of growing-season insolation. The corrected insolation value for 1956 is used in Fig. 1 and in all calculations of the ratio of insolation and production of the vegetation. The 50% reduction may be excessive during April, May and June when the total insolation at ground surface is reduced due to increased cloudiness. Clouds reduce the amount of ultraviolet and infrared radiation because of diffusion and absorption by water molecules (Terrien, Truffaut & Carles 1957) and it might be anticipated that under clouds more than 50% of the insolation at the ground would be usable by the plants.

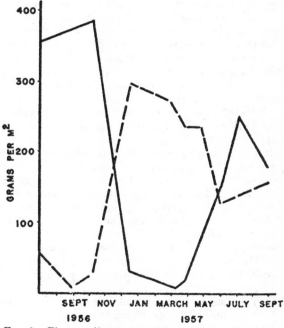

FIG. 2. The standing crops of living (solid line) and dead (broken line) vegetation by months in 1956 and 1957.

Dynamics of the Vegetation

The production of the vegetation can be separated into two different components: (1) the production of the plant tops and the root biomass over the growing season, and (2) the photosynthate which is lost to consumption by animals and to respiration of the plant biomass. The production of tops and roots plus the material eaten by animals comprise the net production; the inclusion of the respiration yields the gross production of the vegetation. In measurements of net production by the harvest method, consumption of green plant material by animals often is not included in the estimate of total net production. Here, food consumption by the mouse population is added to the production of the roots and tops, while food consumption by insects and other herbivores is not estimated or included.

The dry weight standing crop of vegetation (Fig. 2) shows a typical cycle of growth, death, and decay of vegetation. The grass-herb ratio shows cyclical fluctuations; grasses predominate in fall and winter, with a tendency toward equality in midsummer. A slight change in caloric content of the vegetation also occurred seasonally (Tables 11, 12). The peak aboveground standing crop was 385 gm per m^2 (3.85×10^6 gm per ha) in 1956 and 251 gm per m^2 (2.51×10^6 gm per ha) in 1957, these values being accepted as minimum estimates of production. The average caloric value of green vegetation was 4.08 Cal per gm dry weight (average of the values in Table 2).

Standing crop of roots was measured three times during the 1957 growing season. The initial standing crop (1493 gm per m^2, 15 cm deep) was measured on April 13, 1957 when the vegetation was beginning spring growth. The second measurement was made on July 11, when the root standing crop was 1805 gm per m^2. The peak standing crop was 2516 gm per m^2 on September 29, 1957. The difference between the peak and initial standing crop approximated the organic matter synthesized and stored in the roots above 15 cm depth over the growing season (1023 gm per m^2) in 1957. It was assumed that this rate was also applicable to the 1956 season. The caloric value of the roots, determined for a sample collected in August 1956 was 3.30 Cal per gm dry weight (Table 12).

The amount of live vegetation calculated to have been consumed by mice during the growing seasons in 1956 and 1957 is shown in Table 13.

Data on respiration of the vegetation were collected during four nights in the summer of 1957. The CO_2 released, O_2 consumed, RQ, and Cal used per

TABLE 11. Above-ground standing crop of living and dead vegetation, and average caloric value of green grasses and herbs per square meter plot, and living grass-herb ratio.

Date Collected	Number of Plots	Standing Crop Live		(Grams Dry Weight) Dead		Caloric Values Cal/gram Dry Weight	Grass:Herbs (Weight)
		Mean	SD	Mean	SD		
7/21/56	2	—	—	—	—	4.12	—
8/ 8/56	10	358.0	±49.2	55.6	±30.4	4.12	3.5:1
9/24/56	16	372.4	±81.2	8.0	—	4.30	9.1:1
11/ 2/56	20	385.2	±91.2	28.0	—	4.17	49.2:1
2/ 9/57	10	31.2	±12.8	300.0	±60.0	—	—
3/22/57	15	8.0	±2.4	274.0	±75.2	—	—
4/24/57	14	18.4	±6.8	234.8	±97.2	3.99	16.0:1
5/22/57	14	67.6	±12.8	236.0	±66.4	3.99	6.9:1
7/ 1/57	15	147.2	±49.6	126.8	±47.6	3.90	2.6:1
8/ 5/57	15	250.8	±71.2	140.4	±42.8	—	1.4:1
9/29/57	14	184.0	±71.2	164.0	±42.4	—	—

TABLE 12. The caloric value per gram of oven-dried plant tissue for various plant species collected during the study.

Date Collected	Species	Number of Samples	Cal/gm (Aver.)	S.D.
7/22/56	Poa compressa	5	4.12	±.08
8/ 8/56	Poa compressa	9	4.18	±.20
9/24/56	Poa compressa	14	4.31	±.45
11/ 2/56	Poa compressa	15	4.18	±.08
5/22/57	Poa compressa	3	4.02	±.05
7/ 1/57	Poa compressa	3	3.99	±.11
8/ 8/56	Linaria vulgaris	5	4.28	±.07
9/24/56	Linaria vulgaris	5	4.34	±.10
8/ 8/56	Daucus Carota	5	3.92	±.45
8/ 8/56	Cirsium arvense	6	3.93	±.22
8/ 8/56	Trifolium repens	2	4.09	±.20
8/ 8/56	Verbascum Thapsus	3	3.98	±.14
8/ 8/56	Plantago spp.	2	3.79	±.09
8/ 8/56	Dead grass	6	3.91	±.05
2/ 9/57	Dead grass	15	4.25	±.06
4/24/57	Grass and herbs	3	3.99	±.10
5/22/57	Herbs combined	3	3.97	±.12
7/ 1/57	Herbs combined	3	3.81	±.12
8/ 8/56	Roots	6	3.30	±.20

TABLE 13. Food consumption of *Microtus* populations during the 1956 and 1957 growing seasons.

Date	Interval Days	Density Mice Per Hectare	Individual Mean Weight Grams	Population Biomass Grams	Consumption Grams Per Hectare
1956					
April 1.......	52	5.0	29	145	1,056
May 22.......	63	5.2	29	151	1,329
July 24.......	42	9.6	29	278	1,634
Sept. 4.......	35	16.5	29	479	2,349
Oct. 9.......	22	21.2	31	657	2,024
Total......					8,392
1957					
April 1.......	53	64.2[1]	32	2054	15,243
May 23.......	26	49.8	26	1295	4,714
June 18.......	36	65.4	28	1831	9,227
July 24.......	61	61.4	30	1842	15,732
Sept. 23.......	38	111.2	26	2891	15,379
Total......					60,295

[1] Average of March and April population estimates.

gm of plant per night are shown in Table 14. The product of the rate of respiration in Cal, the mean standing crop of vegetation and the number of days between measurements of respiration yielded the Cal used in night respiration by the vegetation over the growing season, 146.2 Cal per m^2 plot or 1.46×10^6 Cal per ha (Table 15).

TABLE 14. Night respiration of old field vegetation.

Date	CO_2 Produced cc./night	O_2 Consumed cc./night	RQ	Calories per cc. Oxygen	Grams Vegetation[1]	Calories per Gram per Night
5/17/57	1.96	3.50	.56	.0045	.398	.041
5/25/57	1.36	1.91	.72	.0047	.589	.016
6/13/57	1.17	1.20	.97	.0050	.866	.007
7/20/57	.90	.68	1.31	.0053	1.905	.002

[1] Grams of vegetation on area covered by cans (86.6 cm²) derived from the graph of standing crop of vegetation (Fig. 1).

Thomas & Hill (1949) in their field studies on the respiration of alfalfa showed that night respiration of the tops was approximately one-half the day respiration and that the root respiration was about equal to the combined day and night top respiration. If we assume that these findings can be applied to the old-field vegetation, the day respiration of the tops would equal 2.92×10^6 Cal per ha, and respiration of the roots, 4.38×10^6 Cal per ha. Total respiration of the entire plant biomass would be approximately 8.76×10^6 Cal per ha or 15% of the total assimilation. This estimate is slightly less than those of Transeau (1926) and Thomas & Hill (1949). These workers suggest that respiration of the plant biomass amounts to 25-35% of the total production. It is not known if the discrepancy between estimates made in this study and those by Transeau and Thomas & Hill is due to a diffusion of gases between the soil and air under the cans, or if it represents a real difference between the respiration of natural vegetation and cultivated crop plants.

In 1956 and 1957 the production was made up of the following components:

	1956		1957	
Production of tops	3.85×10^6gm/ha		2.51×10^6gm/ha	
Root production	$10.23 \times$	"	$10.23 \times$	"
Consumed by *Microtus*	$.01 \times$	"	$.06 \times$	"
Weight net production	$14.09 \times$	"	$12.80 \times$	"
Caloric net production	49.51×10^6Cal/ha		44.25×10^6Cal/ha	
Respiration	$8.76 \times$	"	$8.76 \times$	"
Caloric gross production	$58.27 \times$	"	$53.01 \times$	"

THE DYNAMICS OF THE *Microtus* POPULATION

The energy dynamics of the Microtus population were separated into several components: (1) the tissue production, (2) energy expense of respiration, and (3) the intake of energy through foods. The estimate of tissue production was, in turn, based on determinations of the standing crop of mice, production of young, and rate of growth of mice in all age cate-

TABLE 15. Night respiration of the vegetation biomass during growing season.

Date	Interval Days	Mean Standing Crop Vegetation Grams	Respiration Rate Cal/gm/night	Respiration Loss/m^2 Cal.
April 1 to May 21	51	6.1	.041	50.4
May 22 to June 2	12	16.7	.016	12.4
June 3 to July 2	30	27.6	.007	23.2
July 3 to Nov. 1	121	62.2	.002	60.2
Total				146.2

gories. To relate *Microtus* to the vegetation base of the food chain, determinations of consumption and digestibility of foods were used to estimate the percentage of the available food consumed by the mice, and the percentage of the consumed food used in metabolic processes and stored as tissue production. Finally, each separate component was brought together in Fig. 1 to show the entire exchange through the *Microtus* population.

Standing Crop of *Microtus*

The standing crop of *Microtus* showed some unexpected variations during the investigation (Table 1). The population at the beginning of the study was at a very low level (5.2 mice per hectare). In fact, following the first trapping program in May 1956, an attempt was made to relocate the study on another area with a higher population of *Microtus*. It was not known whether this "low" was the result of adverse weather, heavy predation, or "cyclical behavior."

The estimated peak population (139 mice per hectare) determined in this study could not be considered unusual for the species. Other workers have arrived at greater estimates of population density for *Microtus pennsylvanicus*, 291 per hectare (Bole, 1939), 395-567 per hectare (Hamilton, 1937), and 165 per hectare (Townsend, 1935). The months in which the peak density occurred was unusual. Hamilton (1937), Martin (1956) and others showed that *Microtus* generally reach a peak population in the fall (September to November) after which the population decreases until breeding begins again in the spring. Linduska (1950), on the contrary, found that the annual peak population at Rose Lake, Michigan, approximately seven miles north of this study, occurred in January and February, as in the present study. Linduska was unable to explain the difference between his results and those of Hamilton, but

TABLE 16. Snowfall and new captures of *Microtus* in three winter months compared with a typical summer month.

Day of Trapping	November		January		February		July	
	Snow[1]	Captures	Snow	Captures	Snow	Captures	Snow	Captures
1	.00	2	.00	1	.00	3	.00	4
2	.00	3	.05	1	.00	4	.00	11
3	.00	4	.01	2	.10	10	.00	4
4	.00	3	.28	8	.00	9	.00	6
5	.10	8	.17	2	.00	3	.00	9
6	.50	2	.01	1	.00	7	.00	3
7	T	0	T	2	T	1	.00	2
8			.85	11				
9			.05	0				

[1] Snowfall or sleet in inches, taken from U. S. Department of Commerce, Local Climatological Date for East Lansing, Michigan.

suggested that the winter peak may be a local adaptation to "xerophytic" conditions.

A more detailed knowledge of the population density and topography of the study area enabled the writer to suggest another explanation for the winter population high. A region of dense cover, composed primarily of bluegrass sod, occurred in the center of the trap area. This dense cover may have served as a place of refuge for the mice during alternately wet and freezing weather occurring in January, February, and March. The number of new captures per day of trapping was correlated with snowfall (Table 16) suggesting that snow storms stimulated increased movement of the mice. Since in November and January the increased captures of new animals occurred late in the trapping period, it was thought that these captures represented new animals moving into the trap area rather than movement of the resident population. The movements immediately before and during snowfall were considered to be due to a migration of mice from upland areas into the areas with heavier cover, resulting in an increased population on the study area during the storm periods and possibly throughout the winter.

The fraction of animals captured in common to both lines decreased in January, February, and March (Table 1). This seasonal variation may have been the result of decreased size of the home range of the mice and consequent shorter daily movements. These in turn, may have resulted from increased density of mice or from some characteristic of winter weather acting on mouse behavior. Further information is needed before the cause for the decrease in number of captures in common in the winter can be established.

The observed fraction of animals captured in common to both lines for January (4.5%) was lower

TABLE 17. The wet weight, dry weight, and the average caloric value per gram dry *Microtus* tissue determined for four male mice (standard deviation in parenthesis).

Individual	Age	Live Weight	Dry Weight	Caloric Value of Tissue
1	adult	39.1	11.9	4.49 ($\pm$ .21)
2	adult	24.5	7.0	4.67 ($\pm$ .25)
3	adult	28.0	8.1	4.63 ($\pm$ .26)
4	juvenile	10.0	2.9	4.82 ($\pm$ .07)
average (pooled data)				4.65 ($\pm$ .21)

than that fraction used in Table 1. The January trapping period was interrupted by heavy snowfall and it was possible to run the trap lines only 3 days in one period and 4 days in another. If the unusually low fraction .045 is used to determine population density a very high population estimate (256 per ha) results. It was thought that this high a density was unlikely, since it would require a six-fold increase in the population in two months. Therefore, the fractions of common captures for the other winter months (February and March) were averaged and this average was used to estimate the January population.

Caloric Value of Microtus Tissue

Since the objective of this study was to determine the energy transfer and losses between the levels of the food chain, it was necessary to convert the production data, calculated initially in terms of weight, into Calories (Table 17). In the process of preparing mice for calorific analysis, the mice lost approximately 71% of their body weight. Since mouse production figures were computed in terms of live weight, the average caloric value per gm of mouse tissue had to be converted from dry weight (4.65 Cal per gm) to live weight (1.37 Cal per gm).

Mortality or Emigration of the Mice

Mortality and emigration both result in the disappearance of mice and are considered collectively in this report. When an animal was not caught again, it was impossible to determine whether it had died or had moved out of the trapping area. In a few instances animals were trapped in one month and not retrapped until several months later. Where these animals resided in the intervening period is unknown, but it is here assumed that they were on the trap area.

The mortality of juveniles and adults was greatest immediately after the peak population was reached in February (Table 1). This peak was possibly correlated with periodic inundation of the low portions of the study area in February and March. As was mentioned previously this population decrease was expected to occur in December but may have been postponed by an immigration of mice into the trap area in January and February. Because of the fire on the study area in March, a measure of the mortality rate was unavailable for February. Mortality was assumed to be 55% in this month (based on the difference between the population estimates for February and March on the two adjacent areas).

Production of Young by Adults

The potential number of young produced by the population showed a consistent increase from 24.8 mice per ha in May to 137.9 mice per ha in November 1956, and from 15.2 mice per ha in March to 79.3 mice per ha in April 1957 (Table 3). During January and February no females were judged to be in breeding condition and it was assumed that no breeding occurred. This assumption is consistent with the findings of Hamilton (1937). During the first four months of the study data on the breeding condition of the females were not collected. The pregnancy rate for these periods was later assumed to be approximately 90%.

The potential natality suggests the maximum possible number of young which could be produced by the population, and may be the source of most of the population increase and of replacements for adults disappearing from the population. The number of young entering the trap-susceptible population was assumed to be equal to the number of adults and juveniles disappearing between trapping periods. These replacement young also showed a consistent increase from spring to fall (Table 18) and in all but March were fewer in number than the potential production of young.

Production of Tissue

Production of tissue in the mouse population was calculated for each trapping period from a knowledge of the average population biomass and the observed rates of growth. This calculation was carried out separately for the various weight classes, the production of tissue being estimated by methods described earlier. For each class the average biomass between times of trapping was estimated as the mean value of the corresponding population biomass determinations made at each trapping. The rates of growth were determined from weights of individual mice recaptured in two consecutive trapping periods.

On occasion certain weight classes, while obviously contributing to the population biomass, were not represented by recaptured animals, and hence, no growth rates were available for these classes. In Table 2 these instances are distinguished. Since it was known that some of the classes not represented by recaptured mice contributed to the production of tissue, it was necessary to estimate appropriate rates of growth for weight classes known to be producing tissue.

To estimate missing growth rates, use was made of the fact that growth rates decreased progressively with increasing weight, in each time period. This

TABLE 18. Population dynamics of nestling young on one hectare.

Date	Potential Production Numbers	Replacements[1] Numbers	Survival Rate Per Cent	Mean Biomass of Young Grams	Mortality		Immigration Numbers
					Grams	Calories	
May	24.8	14.5	58	144	78	107	—
July	28.5	12.7	45	135	108	148	—
Sept.	34.8	18.3	53	180	114	156	—
Oct.	64.5	38.0	59	333	176	241	—
Nov.	137.9	63.8	46	893	693	949	—
Jan.	0.0	0.0	—	—	—	—	—
Feb.	0.0	0.0	—	—	—	—	96.0
March	15.2	27.4	100	100	0	0	13.4
April	79.3	44.1	56	361	209	286	—

[1] Number of young replacing adults disappearing through death or migration.

fact is obvious in Table 2 where the proportional weight changes, unadjusted for length of interval, show that over the year the average weight change for mice in the 11-20 gm class exceeded the average change for the 21-30 gm class by a factor of 0.66. Similarly, the 21-30 gm mice exceeded the 31-40 gm animals by 0.30. For those classes for which information was lacking on growth rates, as detailed above, substitute values were approximated by adding the above average increment in proportionate growth to the observed proportionate change for the next heaviest weight class. For example, in October 1956, 0.66 was added to the value of 1.20 observed for the 21-30 gm mice to estimate the missing value for the 11-20 gm animals.

In every instance survivors in the 41-50 gm class lost weight (Table 2). Hamilton (1941) suggests that the heavier adult mice lose weight only in the winter, but these data indicate that weight loss is characteristic of the 41-50 gm weight class throughout the year. This weight loss does not appear to be correlated with senility because in no instance did a heavy animal showing a weight loss disappear in the trapping period immediately following the loss in weight. Over the winter months of January and February mice in most other weight classes also lost weight. These losses probably represent the exhaustion of body fat stored over the fall months.

The weight distribution graph (Fig. 3) reflected the internal dynamics of the mouse population. Since weight may be considered a rough criterion of age, it would be expected that the greatest number of mice would fall in the lightest weight classes. However, no mice were caught in the 1-10 gm weight class, except in May. Whitmoyer (1956) found in his study of the growth rate of laboratory-raised *Microtus pennsylvanicus* that the eyes of all young were open at 11 days of age, at an average weight of 9 gm. Hamilton (1941) also showed that young *M. pennsylvanicus* did not leave the nest until they were 10 gm in weight, at 9-13 days of age. Therefore, it was assumed that nestling young did not leave the nest before they attained a weight of 10 gm, and that the probability of capturing an animal weighing less than 10 gm was very small.

Blair (1948) states that small meadow mice old enough to leave the nest may be caught by live traps. If the probability of capture were the same for all weight classes, in those months in which reproduction occurred we would expect the highest number of captures to be in the 11-20 gm class. However the 11-20 gm class in each month except one (May) had

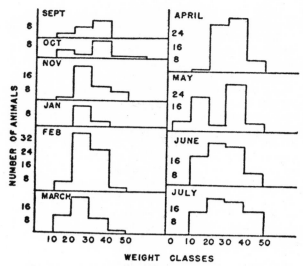

FIG. 3. The number of animals per weight class captured during each trapping period.

fewer mice than did the 21-30 gm class (Fig. 3). The home range of very young mice may be smaller than that of adults, resulting in a lower probability of capture of juvenile mice and a proportionately lower representation of this class of mice in the data. A second explanation for fewer animals in the 11-20 gm class than expected may be that the rapid growth of these mice shortens the period of exposure to trapping for individuals in this weight class. Hamilton (1941), observing the lower proportion of lightweight mice in his trapping data, suggested that exceptionally heavy mortality in this weight class may be further cause for the phenomenon. Whatever the cause, if the probability of capture for the 11-20 gm mice is less than for the 21 gm and heavier animals, the determination of the biomass of the population based on live-trapping data might underestimate the contribution of the 11-20 gm weight class.

The growth rates of nestling young (Table 4) were generally quite similar over the year since the growth rates were calculated from a constant birth weight of 3 gm and the relatively constant weight of the lightest juvenile captured in the traps. Since the nestling young had the highest growth rates, they contributed a larger share of the tissue growth or production over the period of study (Table 4) than did the adults and juveniles (Table 2).

Immigration into the Study Area

In January and March the potential production was insufficient to account for the increase in the population and/or the mortality of adults and juveniles. To account for the discrepancy between the potential production and the number of young entering the trap-susceptible population, it was assumed that mice migrated into the study area. In March this type of immigration was of a relatively minor nature, amounting to only 21% of the mean adult population. In January, however, the immigration was of greater importance since the population almost doubled between January and February (Table 1). There were no young produced in January; therefore, all of this increase must have been due to immigration. This assumption is supported by the previous information of the winter movements of mice immediately before or during snow storms (Table 16), which were interpreted to mean a migration of mice into the study area.

Theoretically, since the population increased in size over the period of study, the energy lost to mortality should not exceed that appearing in tissue production. However, Table 19 shows that mortality exceeded production in every period studied. Since the standing crop increased over the year by 1569 Cal (Table 1), production plus immigration must have exceeded mortality plus emigration by this amount. These last two processes exceeded production alone by approximately 12,000 Cal (Table 19).

TABLE 19. Tissue production, mortality, and respiration of mouse population in calories per hectare.

Period	PRODUCTION		MORTALITY		RESPIRATION	
	Entire Period	Per Day	Entire Period	Per Day	Entire Period	Per Day
May to July	331	5	500	8	4,871	76
July to Sept.	309	8	380	9	5,007	119
Sept. to Oct.	488	14	695	20	6,808	195
Oct. to Nov.	787	20	951	24	10,578	271
Nov. to Jan.	1742	29	2,111	35	33,245	575
Jan. to Feb.	—	—	1,245	36	27,055	773
Feb. to March	—	—	3,284	122	37,544	1391
March to April	727	20	937	26	27,710	770
April to May	786	28	2,217	79	17,059	609
Total	5170		17,200		169,877	

The conclusion follows that immigration (termed import in Fig. 1) must have contributed approximately 13,500 Cal. The estimates of immigration in Table 18 account only for the difference in calculated production of young and the young needed to replace trap-susceptible adults and juveniles. These estimates are included in the above calculations.

Energy leaving one population of mice by emigration joins another by immigration. Over all populations, these gains and losses must balance, just as production of all populations must equal the sum of mortality and population change. In this particular instance, immigration appears to have involved over twice as much energy transfer as production within the population itself. It is not known whether this population, sustaining a heavy population pressure as calculated from the weasels alone, represents a "sink" with energy flowing only inward, or whether it may approach an energy equilibrium, with the weasel and other predators being in equilibrium with the local production and there being a sizable exportation of energy from the population.

Immigration adds energy to a population, increasing the biomass, but not by any process defined here as production. Production of these animals took place elsewhere, and hence, cannot be credited to the local population. On the other hand, energy losses from the respiration of immigrants may have considerable influence on the energy balance in the community. Inspection of Table 1 and 19 will show that immigration increased the biomass greatly during January and February, accounting for a large part of the high population metabolism or respiration..

Respiration of the Mouse Population

An estimate of the minimum number of Calories used by the mouse population in respiration was obtained from the fasting metabolic rate of *Microtus*. These rates showed very little variation seasonally (Table 5). Although an analysis of variance in the metabolic rates of individual groups in each calorimeter jar showed that no significant differences in the rates between seasons or within experiments, seasonal variations are to be expected. These were probably obscured by confining the mice in the laboratory at a constant temperature while measurements were being made. The average metabolic rate of 10 Cal per 24 hrs per mouse was used as the minimum metabolic constant to determine the minimum respiration of the mouse biomass (Table 20).

The average metabolic rate determined in the present study, 2.75 cc O_2 per gm per hr, compares favorably with the determinations on *Microtus pennsylvanicus* made by Pearson (1947), 1.9—2.8 cc/gm/-hr, and Morrison (1948), 1.8—2.8 cc/gm/hr. Hat-

TABLE 20. Respiration of the mouse population.

Date	Respiration of Young	Respiration of Adults
	Calories per Hectare	
May	1,575	3,296
July	975	4,032
Sept.	1,050	5,758
Oct.	2,329	8,249
Nov.	7,381	25,864
Jan.	—	27,055
Feb.	—	37,544
March	620	27,090
April	2,177	14,882
Total	16,107	153,770

field (1939) studied metabolism in *Microtus californicus* but arrived at a much higher figure of metabolism, 5.2 cc/gm/hr. Pearson suggested that the higher metabolic rates obtained by Hatfield were due to greater activity during the periods of measurements.

The metabolic rate determined by this study was a minimal rate and when used in calculations of energy loss by respiration underestimated the actual loss. The average metabolic rate of *Microtus* in the wild is unknown. Morrison (1948) gives some indication of what this average rate would be; in his studies, the average metabolic rate of Microtus held in the calorimeter for 24 hrs, with food and water provided, was 18-42% above the minimum rate of metabolism.

Summary of *Microtus* Energy Assimilation

The summary of the energy assimilation of the *Microtus* population (Table 19) shows the assimilation of a population changing from a low density to a higher density. With the exception of January and February, when growth of young and adult mice did not occur, tissue production rose continuously during the spring, summer, and fall of 1956, and the spring of 1957. Mortality, including the disappearance of mice from the population by death, predation, or emigration, also showed a smooth rise, with a disproportionate increase in February. Respiration was proportional to the population biomass and increased until the population drop in March. Following this, the energy loss due to respiration began a second increase. Laboratory metabolic measurements allowed no estimate to be made of expected seasonal changes in the respiration energy cost.

The respiration loss, though a minimal estimate,

accounted for by far the greatest amount of energy used by the *Microtus* population. Respiration accounted for 68% of the energy passing through the mouse population yearly. In comparison, the mean biomass of mice (1900 Cal per ha) represented a storage at any one time of only 1.0% of the annual energy consumption; the maximum biomass (Table 1) was only 3% and the minimum biomass 0.1% of the annual energy consumption.

Digestibility of Food

The amount of food ingested and its energy value may be understood in relation to the metabolism of the mouse population only through some knowledge of the digestive processes and their efficiency. The gross energy in the food minus the energy in the feces equals the digestible energy (Brody 1945). The feces include secretions and cells sloughing into the digestive tract. Thus the digestible energy indicates only the excess of food energy taken into the blood stream over excretion into the gut.

The digestibility of the food was investigated briefly in the laboratory, with results from three animals as indicated in Table 21. Although requiring confirmation, they indicate an exceptionally high digestibility for alfalfa in *Microtus* (90% of the gross energy as against the figure of 50% given by Morrison

TABLE 21. Digestibility of laboratory diets.

Day	Caloric Intake in Food	Caloric Loss in Feces	Undigested (per cent)
Alfalfa diet—2 female *Microtus*			
1	128.7	20.2	15.7
2	183.1	15.2	8.3
3	83.3	12.9	15.5
4	143.5	12.0	9.0
5	176.1	12.4	7.0
average (pooled data)			10.2 (± 4.2)[1]
Standard diet—1 male *Microtus*			
1	26.1	4.3	16.5
2	21.3	4.3	20.2
3	28.9	4.4	15.2
4	21.5	4.3	20.0
5	23.3	4.2	18.0
average (pooled data)			17.8 (± 2.2)

[1] Standard deviation in parenthesis.

1949 for cattle and sheep.) Further, with the diet of lettuce, carrots, and oatmeal, 82% of the gross energy was digested, equalling the presumed digestive efficiency of humans with the same diet, as calculated from the tables of Merrill & Watt (1955).

Further work in this direction is needed to understand the trophic ecology of *Microtus*, as well as other small mammals. Attempts should be made to determine the digestibility for various single dietary components and for complete diets. The qualitative composition of the diet also requires investigation, as suggested in the following section.

Consumption of Food

The amount of food consumed, considered with digestibility, constitutes the trophic levy upon the environment by the population. Food consumption may be estimated in several ways; here it has been done by laboratory means (Table 7) and by examination of the stomach contents of wild mice (Table 22).

TABLE 22. Weight of stomach contents of snap-trapped mice.

Season	Number Weighed	Mean Wgt Mice Grams	Mean Weight of Contents Grams	Standard Deviation Grams	Calculated[1] Daily Food Consumption Grams
Fall	28	28	1.15	±.83	23.0
Winter	27	28	1.35	±.68	27.0
Spring	21	35	1.17	±.59	23.4
Summer	24	27	1.31	±.68	26.2
average					24.9

[1] Under the assumption that the mean weight of stomach contents represents one-half the stomach capacity and that the mice have 10 feeding periods per day.

Food consumption of *Microtus* has long been of interest to investigators, partly for economic reasons. Bailey (1924) fed *Microtus* a diet of clover, cantaloupe, grain, and seeds and found that they consumed 55% of their body weight daily. Regnier & Pussard (1926) obtained similar results with *Microtus arvalis* on a mixed diet of oat seeds, oat stems, and mangolds. In an experiment on which *Microtus* were fed a diet of dry feed (rolled oats, dry skim milk, dry meat, and seeds), Hatfield (1935) found an average consumption of 3.48 gm of food per mouse per day.

In the present study the mice on the standard diet consumed more fresh food than did the mice on the alfalfa diet. When dry weight of the food and caloric value was considered the reverse was true. It appears from these results and with comparison with Hatfield's findings that inclusion of dry foods in the diet reduced food consumption on a weight basis. The dry food materials (oatmeal, seed, etc.) have a much higher caloric value than fresh leafy foods such as lettuce and alfalfa, and animals on a fresh diet of succulent foods might have to consume a greater quantity of food to satisfy their energy requirements. In the present study, the mice on the alfalfa diet consumed 61% of their body weight daily and those on the "standard" diet consumed 86%. In both instances, the food consumption as a percentage of body weight was greater than that found by Bailey (1924) and Regnier & Pussard (1926).

Since blue grass appeared to be the dominant food plant of the environment in the study, attempts were made to maintain *Microtus* on a diet of fresh-cut, mature blue grass, with water available. In two attempts most of the animals lost considerable weight, or died, as indicated below:

Trial	Number Mice	Number Dying	Number Losing Weight	Number Gaining Weight
1	6	2	2	2
2	4	2	2	0

Dice (1922) was able to maintain *Microtus ochrogaster* on blue grass; possibly he fed immature grass or grass sod which might have a higher nutritive value. The energy content of the blue grass was 4.13 Cal per gm dry weight, which was closely similar to the caloric value of alfalfa (4.12 Cal per gm); however, the protein content of mature blue grass (6.6%, Morrison 1949) is much lower than that of alfalfa (14.8%, Morrison 1949). Lack of protein may be a cause of my failure to maintain *Microtus* on a blue grass diet. Regnier & Pussard (1926) found *Microtus arvalis* ate meat readily, consuming other voles and insects (Carabidae). They suggest that this consumption of protein might influence the numbers of mice during plague years. During one experiment in the present study, one *Microtus* ate 15 large grasshoppers within 24 hours. *Microtus* may thus supplement a low protein diet with insects or other high protein foods.

The stomach contents of animals taken by snap-traps reveal the proportional composition of the diet, assuming all components to be digested at the same rate (Table 6). To infer food consumption from the information on the volume of food in stomachs further, information on the rate of stomach clearance through digestion is required. Such information is not available, but one may infer from the observations of Hatfield (1940), Davis (1933), and Pearson (1947) that wild mice characteristically have 8-12 activity periods during a 24-hr day, and that these periods are concerned with feeding activity, to fill a nearly empty stomach.

The quantitative information on stomach contents (Table 22) may be examined, under the assumption of 10 activity periods a day, and a filling of the stomach to twice the mean observed contents at each activity period. The assumption that the mean stomach content equalled half a full stomach is supported not only by theoretical sampling considerations, but also by the observation that the observed stomachs ranged from full to almost empty, with most being "half-full."

The overall estimate of about 25 gm of food eaten per day (.86 gm wet food per gm mouse), for all sizes of capture-susceptible mice, agrees fairly well with the laboratory-determined values of 39 and 28 gm (.86 and .61 gm wet food per gm mouse) for two different diets. It is not known whether the indicated seasonal fluctuations are real or reflect bias from either shifting activity patterns, changing age structure, or sampling variation. However, this method of observation seems to offer a practical, if approximate, method of measuring food consumption.

The differential seasonal consumption of available food materials in Table 6 showed that grass (grass and sedges) was the dominant food at all seasons. Dead vegetation (with the exception of wood) was not found in the stomach slides, therefore, it was assumed that dead vegetation was not used by *Microtus*. Since the clip quadrat used to determine food availability and production of vegetation only sampled grasses and herbs and not mosses, fruit, and other foods, the separate percentages of grass and herbs in the stomachs were also calculated (Table 6).

The food consumption of the population was estimated by multiplying the biomass of the trap-susceptible population by the food consumption (.14 gm dry food per gm of mouse tissue per day or .58 Cal per gm of mouse tissue per day) of captive mice on the standard diet. Food consumption determined with the standard diet was used because it was thought that the standard diet more closely represented the diet of wild mice. Total consumption of vegetation per trap period by the mouse population was 250 × 103 Cal.

The Least Weasel Population

During the study least weasels were captured in 15 live-traps and were tracked on three different days during the winter. The largest number of individuals captured in one trap period was 4 (May 1957). Examination of the entire area on three days in winter (in each instance the morning following a snowfall of the previous day) yielded the tracks of one weasel in December, two in January, and three in February. According to Polderboer (1942) the maximum home range of the least weasel is 2 acres. Since the study area averaged 6.2 acres in size, it was assumed that at least three or four adult weasels could live on the area. Although the first evidence of weasels was not noted until July, 1956, it was assumed that two weasels were present at the beginning of the study in May, 1956.

Burt (1948) states that two litters of young are born per year. Litter size ranges from 4 to 10 (Burt 1948) and averages 5 (Hall 1951). If 2 litters of 5 young were produced by the weasels over the year, it is estimated that approximately 12 weasels were present on the area in September, 1956. This population of 12 animals decreased to 6 animals in May 1957. The mean weight of the young dying in the period August to December was estimated at 15 gm, and in January to May at 35 gm. Although the population values may appear unusually high since the least weasel has been considered a rarity in Michigan (Hatt 1940), the evidence available supports these estimates.

Under the assumption that the population of least weasels followed the model developed in the section on methods and further elaborated in the above introductory paragraphs, the production of tissue by the weasels remained rather steady throughout the year (Table 8). In the summer, production of tissue was due to growth of both adults and young. In the fall and winter, the population of young, decreasing from 10 to approximately 5 animals, contributed all of the growth in this period. In the late winter and spring, the population of young, decreasing from 5 to 4 animals, again furnished all of the growth of tissue.

The respiration energy loss of the weasel population, based on an average minimum rate of O_2 consumption of 1.61 cc per gm per hr (Morrison 1957), increased over the year of study (Table 8). As observed with *Microtus*, the energy used in respiration of the weasel biomass was considerably greater than that involved in tissue production.

The laboratory feeding experiments were used to evaluate the role of the least weasel as a predator on *Microtus*. *Microtus* were assumed to be the sole food used by the weasel (Hatt 1940). In the laboratory experiments, the captive weasel consumed either 15.1 gm of white mice or 14.7 gm of *Microtus* per day (Table 9). Llewellyn (1942) found a similar rate of consumption in studies with a 32-gm weasel, i.e. 19.7

gm of mice per day. An average food consumption of 15 gm of mouse tissue per day was assumed to represent the true food consumption of adult weasels over the year. The young weasels, like other mammals (Morrison 1949), would probably use less food per day than the adults, and the daily food consumption of young was estimated to be 5 gm per day from May to August and 8 gm per day from August to December. Using these constants as a basis for calculating true food consumption, the effect of the weasel on the *Microtus* population was estimated. The weasel population consumed 5,824 Cal annually; this consumption was 3.07 times the mean biomass of *Microtus* (1900 Cal) over the year. Since net production of the *Microtus* population totaled only 5,170 Cal per ha annually, the weasel population appears to have required energy in excess of that produced by its principal prey. As previously noted, the production of the *Microtus* population did not allow for the energy imported by the mice which moved into the area. Further, it is possible that the weasel had other sources of food, such as *Blarina* which existed in moderately large numbers on the area, and insects, which were abundant during the summer, or perhaps the calculations of weasel population density here are in error.

The percentage of the energy in the mouse carcasses which was digested by the least weasel was determined in the digestibility experiments. When the weasel was on the *Microtus* diet, he was able to digest 89.9% of the energy in the mouse bodies (Table 9). This rather high efficiency of digestion is comparable with the somewhat lower efficiency of digestion (70-80% of a dry-feed diet, McCay 1949) for the dog.

DISCUSSION

In the transition of energy between two steps of the food chain there are two main pathways by which energy can be lost or diverted from the food chain itself. First, in tracing the energy from one population to the next not all the food organisms will be consumed by the consumer species; some of this energy could be dispersed to another food chain by migration of the food species out of the study area, by consumption of the food species by organisms outside of the food chain, or by death. These are considered to be energy losses of the first order. Second, not all the energy consumed is used in growth or in production of young; some of it is diverted to the maintenance of the organism and some passes through the body unused. These are energy losses of the second order. In energy losses of the first order the energy lost from the food chain is still in a form available for use by other animals. In energy losses of the second order, the loss is primarily heat derived from metabolic processes which is unavailable for further use by the food chain; that passing through the body is, of course, available to various other organisms in the food web.

ENERGY LOSSES OF THE FIRST ORDER

Of the solar energy available to the vegetation over the growing season (one-half of the total incident insolation during the growing season), 1.2% was utilized in the gross production, and 1.1% in the net production of the vegetation (Fig. 1). These figures can be compared with the giant ragweed ecosystem in Oklahoma, 1.2%, and alfalfa growing in experimental plots for 6 months, 3.1% (data converted from tabular material in Odum, 1959—net production was divided by incident energy data for Michigan). Data on the percentage of solar energy utilized by the vegetation of different communities are still too limited to allow any comparison of the efficiencies of a successional community with a stable one. At this time we may say only that terrestrial vegetation of the old field community on this soil, at this site, and for these years appeared to utilize approximately 1% of the available solar energy during the growing season.

The net production of the vegetation can be considered as the energy available to the herbivorous animals in the community. *Microtus* is primarily a herbivore, with animal food appearing only in trace amounts over the year (see Table 6). It was assumed in this study that only the production attributed to the above-ground vegetation could be utilized by *Microtus*. Some of the root biomass was undoubtedly also used by the mice, but no estimate of the extent of root utilization was available. Lantz (1907) and Bailey (1924) state that consumption of roots is relatively unimportant, except during the winter, and roots were not recorded among the stomach contents in the present study. It is assumed here that use of roots was negligible.

Of the energy in the vegetation presumably available to the mice (15.8×10^6 Cal), 1.6% was consumed, with 1.1% utilized by the mice in production and respiration. These percentages assume no loss of vegetation due to cutting of stems and leaves by *Microtus*. The utilization of the energy in the vegetation not consumed by mice was not traced further in this study. Some of this energy was probably diverted through invertebrate food chains. Wolcott

(1937) indicated that insects ate $.94 \times 10^6$ gm (3.76×10^6 Cal) of above-ground vegetation per ha in a pasture in New York over the summer. This level of consumption would amount to 23.9% of the available energy in the old field vegetation of the present study. If these data are correct, insects may be considered as more important herbivores in this old field community than are the meadow mice.

As mentioned previously, the energy in the *Microtus* population (production) available to predators was augmented by an import of energy through immigration of mice into the study area. This immigration was particularly noticeable in the winter and spring of 1957. Of the total energy available in the *Microtus* population (production plus immigration, Fig. 1) the least weasel consumed 31% as food and used 30% in production and respiration; considering only the production of the mice, the weasel consumed over 100%. When the energy consumed by the weasel and the energy retained in the mouse population through an increase in population size from May 1956 to May 1957 was subtracted from the production plus energy imported through immigration, 43% of the energy of the mouse population was unaccounted for. This loss may be attributed to emigration, to death from disease or accident, and degradation through microorganism food chains, or to capture in other predator chains. Some possible predators are *Blarina brevicauda* (Eadie 1952), *Felis domesticus* (Toner 1956, Korschgen 1957), or Owls, Red-tailed hawks, Red-shouldered hawks, and Cooper's hawks (Linduska 1950). All of these predators were seen on the study area during the investigation.

Most of the calculated production of the weasel population could be accounted for by the increase in the size of the population from May 1956 to May 1957. The expansion of the weasel population presumably was directly related to the expansion of the *Microtus* population. Only 10% of the production was not accounted for (Fig. 1). The least weasel may itself serve as food for certain predators, such as the great-horned owl, the barn owl, long-tailed weasel, and domestic cat (Hall 1951), but no information was gathered on mortality of weasels during this investigation.

Energy Losses of the Second Order

The energy losses of the second order, due to respiration (fasting metabolism), nonassimilation of energy in the food, and the energy cost of maintaining the body under normal activity, can be further separated into energy losses available and unavailable to the biosphere. Energy losses available to the biosphere include the energy in fecal matter which is composed primarily of unassimilated food but also contains intestinal secretions and cellular debris. Energy in the feces would serve as the base for food chains of coprophagous organisms. The energy lost to the biosphere as heat derived from animal metabolism can be considered an increase in the positive entropy of the ecosystem.

Respiration loss (respiration energy/energy consumed) determined for each step of the food chain is as follows: Vegetation—15.0%, Mice—68.2%, Least Weasels—93.3%. The respiration coefficient for the vegetation may be underestimated as stated earlier; Transeau (1926) and Thomas & Hill (1949) suggest that it may run as high as 25-30%. The data confirm the statement of Lindeman (1942) that as energy passes through the trophic levels an increasing percentage is lost in respiration. It must be remembered that the percentages cited above for the mice and weasel were both determined as minimum metabolic rates and, therefore, indicate basic differences in the loss due to metabolism and not mere differences in activity. Taking activity into account would presumably act to increase the difference.

The energy consumed but not assimilated by *Microtus* and *Mustela* was measured in the digestion trials. Of the energy consumed, 10-18% was recovered in the feces of *Microtus* and 10% in the feces of *Mustela*.

When these losses were subtracted from the energy loss of the second order for each species, 14-22% of the energy consumed was unaccounted for in *Microtus* and all the energy loss was accounted for in the least weasel. It is highly unlikely that the energy loss in respiration and in the feces comprises the total energy losses in the organism. For instance, energy is also lost in the urine, in fermentation gases, and in specific dynamic action of feeding (energy used in the processes of food utilization). Estimates of the energy loss in the urine are 15% for cattle and 7% for rabbits after fecal losses are subtracted from the gross food energy (Brody 1945: 28), and the loss in specific dynamic action varies from 40% of the intake energy for lean meat, 15% for fat, and 6% for sucrose (Brody 1945: 61).

An additional energy loss not included above is the expense of normal body activity above rest. Very few investigators have been concerned with this maintenance cost according to Brody (1945), although he estimates that this loss is twice the basal metabolic rate. The minimum energy expended when the animal is confined and fasting (fasting metabolism) is known for *Microtus* and *Mustela* and was used to determine the loss of energy in respiration; but the energy used by these animals as they live in their natural environment is completely unknown. This latter energy expense reduces the efficiency of conversion by widening the gap between energy intake and production. The maintenance losses are reflected in the coefficients of production. Plants are the most sedentary and have the highest coefficient, 94.3% (net production/gross production). The weasel is the most active, since it must hunt for its food, and probably has the highest maintenance cost, with a low coefficient, 2.2% (production/energy intake). Possibly the energy expense of hunting by the predator will vary with different densities of the prey population. *Microtus* which is primarily dependent on vegetative material could be expected to have a low cost of maintenance and to display a higher coefficient than that shown in the study (2.1% production/energy intake). If the total energy losses in respiration and in the feces plus hypothetical maintenance cost are added to the production, the energy used totals more than the energy consumed by both the *Microtus* and *Mustela* populations. The reasons for this discrepancy were not determined, but food consumption rates estimated in the laboratory possibly underestimate the true food consumption of the more active animals in the field.

Since the maintenance cost is irreducible, the percentage of energy converted to production will be highest when the birth of new animals and the growth rate of living animals are the highest. This surge of production can most easily occur when sources of high energy food are available. *Microtus* conforms to this model, since the young are born and the rate of growth is highest during periods of greatest growth of the vegetation. It would further increase the year-round efficiency of the population to have the lowest population density during the period of little or no production by plants since at that time the dangers of over-exploitation of the food supply would be greatest. A low density at times of low plant production is the usual observation in field studies of the population dynamics of *Microtus* (Blair 1940, Martin 1956, Greenwald 1957, and others).

Finally, *Microtus* appears, on the basis of energy relationships, to be a relatively unimportant component of the community. Even when the energy consumption of insects (estimated as approximately 24% of the net production) is added to that of the mice, only 25% of the net production of the plants is accounted for. Odum (1959) emphasizes the distinction between the herbivores which eat green plants directly and the delayed feeders which eat dead plant material. Apparently, in this stage of old field succession the major portion of the plant net production is directed through these decomposer food chains.

SUMMARY

The energy dynamics of the perennial grass-herb vegetation—*Microtus pennsylvanicus*—*Mustela rixosa* food chain of the old field community was studied from May 1956 to September 1957.

Solar insolation for the growing season, measured at East Lansing, Michigan, totaled 94.2×10^8 Cal per ha in 1956.

Primary net production by plants was broken down into the following components: (1) production of tops, (2) production of roots, and (3) consumption by *Microtus*. The net production of vegetation for 1956 was 49.5×10^6 Cal per ha and for 1957 was 44.3×10^6 Cal per ha. Respiration of the vegetation during the growing season amounted to approximately 15% of the net production, determined by crude field calorimetry. Gross primary production ranged from 58.27×10^6 Cal per ha in 1956 to 53.01×10^6 Cal per ha in 1957.

Population dynamics and weight changes of the *Microtus* population were studied by live-trapping. Tissue production of young and adult mice was 5,170 Cal per ha per yr. The fasting metabolic rate of mice determined in the laboratory was approximately 10 Cal per mouse per day. Energy lost to respiration equalled 169,877 Cal per ha per yr. The total energy used in growth of the weasel population were 130 Cal per ha per yr and in respiration, 5,434 Cal per ha per yr.

Stomach sample analysis indicated that *Microtus* ate primarily green grass and herbs. Weasels were assumed to feed predominantly on *Microtus*. Total yearly food consumption of the study area as determined from laboratory experiments, was 250,156 Cal per ha for Microtus and 5,284 Cal per ha for *Mustela*.

Of the solar energy available during the growing season, the vegetation used 1.2% in gross production and 1.1% in net production. These results compared favorably with the coefficients for primary production of terrestrial and aquatic communities determined by other workers. Of the energy available to the mice, 1.6% was consumed and 1.1% was utilized in growth and respiration. The weasel population consumed 31% of the energy available to it in the form of *Microtus*, and used 30% of the energy consumed in growth and respiration.

Twenty-one % of the production of the *Microtus* population and only 10% of the weasel production was lost from the food chain. These losses were diverted to other food chains through other predators or through micro-organisms.

Of the energy consumed only a portion was used in production; most of the energy went to respiration or passed through the digestive tract unused. Respiration cost increased from the vegetation level to the carnivore level of the food chain. The gross energy in the experimental diets which was recovered in the feces amounted to 10-18% for *Microtus* and 10% for *Mustela*.

LITERATURE CITED

Bailey, V. 1924. Breeding, feeding and other life habits of meadow mice (Microtus). Jour. Agr. Res. 27: 523-536.

Baker, J. R. & R. M. Ranson. 1932. Factors affecting the breeding of the field mouse (*Microtus agrestis*) Part 1. Light. Proc. Royal Soc. London (B) 110: 313-322.

Baten, W. D. & A. H. Eichmeier. 1951. A summary of weather conditions at East Lansing, Michigan prior to 1950. Mich. State College Ag. Exp. Stat. 63 pp.

Beckwith, S. L. 1954. Ecological succession on abandoned farm lands and its relationship to wildlife management. Ecol. Monog. 24: 349-376.

Blair, W. F. 1940. Home range and population of the meadow vole in southern Michigan. Jour. Wildlife Mangt. 4: 149-161.

———. 1948. Population density, life span, and mortality rates of small mammals in the bluegrass meadow and bluegrass field associations of southern Michigan. Amer. Midland Nat. 40: 395-419.

Bole, B. P., Jr. 1939. The quadrat method of studying small mammal populations. Cleveland Mus. Nat. Hist. Sci. Publ. 5(4): 15-77.

Brody, S. 1945. Bioenergetics and growth. New York: Reinhold Publ. Corp. 1023 pp.

Burt, W. H. 1948. The mammals of Michigan. Ann Arbor: Univ. Mich. Press. 288 pp.

Clarke, G. L. 1946. Dynamics of production in a marine area. Ecol. Monog. 16: 321-335.

Clarke, G. L., W. T. Edmondson & W. E. Ricker. 1946. Mathematical formulation of biological productivity. Ecol. Monog. 16: 336-337.

Crabb, G. A., Jr. 1950a. Solar radiation investigations in Michigan. Mich. Agr. Expt. Sta. Tech. Bull. 222. 153 pp.

———. 1950b. The normal pattern of solar radiation

at East Lansing, Michigan. Mich. Acad. Sci., Arts, and Letters 36: 173-176.

Daubenmire, R. F. 1947. Plants and Environment. New York: J. C. Wiley Co.

Davis, D. E. 1956. Manual for analysis of rodent populations. Ann Arbor, Mich.: Edwards Bros. 82 pp.

Davis, D. H. S. 1933. Rhythmic activity in the short-tailed vole, *Microtus*. Jour. Anim. Ecol. 2: 232-238.

Dice, L. R. 1922. Some factors affecting the distribution of the prairie vole, forest deer mouse, and prairie deer mouse. Ecology 3: 29-47.

Eadie, W. R. 1952. Shrew predation and vole populations on a localized area. Jour. Mammal. 33: 185-189.

Evans, F. C. & S. A. Cain. 1952. Preliminary studies on the vegetation of an old-field community in southeastern Michigan. Contrib. Lab. Vert. Biol. Univ. Mich. 51: 1-17.

Fernald, M. L. 1950. Gray's Manual of Botany. 8th Ed. New York: Amer. Book Co.

Greenwald, G. S. 1957. Reproduction in a coastal California population of the field mouse *Microtus californicus*. Calif. Univ. Pubs. Zool. 54: 421-446.

Hall, E. R. 1951. American weasels. Kans. Univ. Pubs. Mus. Nat. Hist. 4: 1-466.

Hamilton, W. J., Jr. 1937. The biology of microtine cycles. Jour. Agr. Res. 54: 779-790.

———. 1941. Reproduction of the field mouse, *Microtus pennsylvanicus* (Ord). Cornell Univ. Agr. Expt. Sta. Mem. 237.

Hatfield, D. M. 1935. A natural history study of *Microtus californicus*. Jour. Mammal. 16: 261-271.

———. 1939. Rate of metabolism in *Microtus* and *Peromyscus*. Murrelet 20: 54-56.

———. 1940. Activity and food consumption in *Microtus* and *Peromyscus*. Jour. Mammal. 21: 29-36.

Hatt, R. T. 1930. The biology of the voles of New York. Roosevelt Wildlife Bull. 5(4): 509-623.

———. 1940. The least weasel in Michigan. Jour. Mammal. 21: 412-416.

Korschgen, L. J. 1957. Food habits of coyotes, foxes, house cats, and bobcats in Missouri. Missouri Fish and Game Div. P-R Series No. 15. 63 pp.

Lantz, D. E. 1907. An economic study of field mice. USDA Biol. Surv. Bull. 31: 1-64.

Lindeman, R. L. 1942. The trophic-dynamic aspect of ecology. Ecology 23: 399-418.

Linduska, J. P. 1950. Ecological landuse relationships of small mammals on a Michigan farm. Mich. Dept. Cons. Game Div., Lansing. 144 pp.

Llewellen, L. M. 1942. Notes on the Alleghenian least weasel in Virginia. Jour. Mammal. 23: 439-441.

Martin, E. P. 1956. A population study of the prairie

vole (*Microtus ochrogaster*) in northeastern Kansas. Kans. Univ. Pubs. Mus. Nat. Hist. 8(6): 361-416.

McCay, C. M. 1949. Nutrition of the dog. Ithaca, N.Y.: Comstock. 337 pp.

McLagen, N. F. & M. M. Sheahan. 1950. The measurement of oxygen consumption in small mammals by a closed circuit method. Jour. Endocrin. 6: 456-462.

Merrill, A. L. & B. K. Watt. 1955. Energy value of foods—basis and derivation. USDA Agr. Handbook No. 74. 105 pp.

Morrison, F. B. 1949. Feeds and Feeding. Ithaca, N.Y.: Morrison Publ. Co. 21st Ed. 1207 pp.

Morrison, P. R. 1948. Oxygen consumption in several small wild mammals. Jour. Cell. and Compar. Physiol. 31: 69-96.

Noddack, W. 1937. Der Kohlenstoff im Haushalt der Natur. Ztschr. f. Angew. Chem. 50: 505-510.

Odum, E. P. 1959. Fundamentals of Ecology. 2nd. Ed. Philadelphia: Saunders. 546 pp.

Odum, H. T. 1956. Efficiencies, size of organisms, and community structure. Ecology 37: 592-597.

Park, T. 1946. Some observations on the history and scope of population ecology. Ecol. Monog. 16: 313-320.

Pearsall, W. H. & E. Gorham. 1956. Production ecology. 1. Standing crops of natural vegetation. Oikos 7(2): 193-201.

Pearson, O. P. 1947. The rate of metabolism of some small mammals. Ecology 28: 127-145.

Polderboer, E. B. 1942. Habits of the least weasel (*Mustela rixosa*) in northeastern Iowa. Jour. Mammal. 23: 145-147.

Regnier, R. & R. Pussard. 1926. Le campagnol des champs (*Microtus arvalis* Pallas) et sa destruction. Ann. des Épiphyt. 12(6): 385-535.

Ricker, W. E. 1946. Production and utilization of fish populations. Ecol. Monog. 16: 373-391.

Shively, S. B. & J. E. Weaver. 1939. Amount of underground plant materials in different grassland climates. Nebr. Univ. Conserv. Bull. No. 21. 67 pp.

Tansley, A. G. 1935. The use and abuse of vegetational concepts and terms. Ecology 16: 284-307.

Terrien, J., G. Truffaut, & J. Carles. 1957. Light, vegetation and chlorophyll. New York: Philosoph. Libr. 228 pp.

Thomas, M. D. & G. R. Hill. 1949. Photosynthesis under field conditions. In, Photosynthesis in Plants, edited by J. Franck and W. E. Loomis. Ames: Iowa State Col. Press. 500 pp.

Toner, G. C. 1956. House cat predation on small animals. Jour. Mammal. 37: 119.

Townsend, M. T. 1935. Studies on some of the small mammals of central New York. Roosevelt Wildlife

Ann. 4: 6-120.

Transeau, E. N. 1926. The accumulation of energy by plants. Ohio Jour. Sci. 26: 1-10.

Veatch, J. O., et al. 1941. Soil Survey of Ingham County, Michigan. USDA Soil Survey Series 1933, No. 36. 43 pp.

Whitmoyer, T. F. 1956. A laboratory study of growth rate in young *Microtus pennsylvanicus*. Unpubl. Master's thesis, Mich. State Univ. 62 pp.

Wolcott, G. N. 1937. An animal census of two pastures and a meadow in northern New York. Ecol. Monog. 7: 1-90

Wooster, H. A., Jr. & F. C. Blanck. 1950. Nutritional Data. Pittsburgh: H. J. Heinz Co. 114 pp.

COMMUNITY METABOLISM IN A TEMPERATE COLD SPRING

John M. Teal

INTRODUCTION

The study of community metabolism is one means of making a functional analysis of an ecosystem. Essentially it consists of the study of energy transformation by the organisms of an ecosystem. It provides a measure of the total activity of a community just as a study of individual metabolism does for an individual organism.

The present study of the relatively simple ecosystem of a cold spring was undertaken to provide a more exact measurement of community metabolism than had been available. It should be emphasized, however, that in the present state of our knowledge of community metabolism considerably more assumptions have to be made in order to present a complete picture than would be the case in many other fields.

Studies of community metabolism have been generally made either in terms of energy or of biomass (either as biomass itself or in terms of a portion of the biomass such as protein or fat). The author follows the lead of Macfadyen (1948) in believing energy units to be preferred in studies of community metabolism.

Biomass units are less suitable because there is recirculation of matter in the ecosystem and because the rates of turnover are so different for different sizes and species of organisms. Macfadyen (1948) has shown that confusion often results from the fact that many authors fail to see the distinction between the cycle of matter in a community and the flow of energy through a community. For example, Gerking (1954) states that the variability in quantity of fat in organisms makes calories an unsuitable unit for production studies.

Energy enters the organic world in the form of

sunlight which is absorbed by the green plants and this energy is then used by those plants and by the organisms which feed upon the plants to do their internal and external work. The energy which enters the ecosystem in the form of heat is not usually important. Warm-blooded animals, if their body temperature is maintained by their environment expend less energy in keeping themselves warm. However, this effect is not important for several reasons: (1) the temperature range of all animals is small; (2) most animals are not warm-blooded; (3) related animals which live at and are adapted to different temperatures tend to have a similar rate of metabolism (Bullock 1955); (4) experiments with large domestic animals have shown that the amount of energy saved by environmental heating of the animal's body is negligible.

There are two essential points about the transfer and use of energy that need emphasis. The first is that according to the law of conservation of energy, whatever energy is used by the organisms in doing work will appear as a definite amount of heat which is lost as far as the organisms are concerned. The second is that whatever path the chemical reactions follow, an identical amount of energy is released in the oxidation of a unit amount of an organic compound.

In practice it is not possible actually to measure the increase in heat that results from the organisms' transformation of energy. It is necessary to calculate the calories transformed from data obtained through respiratory rate measurements. This can be done with the aid of the average oxycalorific coefficient determined by Ivlev (1934) since all of the energy that an aerobic organism uses is derived from the oxidation of organic compounds. Allowance must be made for the respiration of that biomass which was produced within one sampling interval and which also died within that interval and so did not appear in any sample (Birch & Clark 1953).

The trophic level concept of Lindeman (1942) has been used in this paper although it has been modified to meet the objections of Ivlev (1945) by considering each important species separately and by dividing the energy flow, for a population that functions to an important extent on more than one trophic level, among the levels concerned. Lindeman's methods of quantifying his trophic level analysis, used also by Dineen (1953), are open to many criticisms (Birch & Clark 1953, Macfadyen 1948) and in this study more accurate methods have been used.

The calculation of various ratios is of value in comparing the energy flow in different species and different communities. The terms in these ratios are defined here as follows: "Assimilation" is the rate of energy assimilation by a population; "Energy Transformed" or "Respiration" is the rate of energy use; "Net Production" is the difference between the previous two. "Gross Production" is used only in reference to primary producers and refers to the energy fixation (Odum 1956).

The ratio, $\frac{\text{net production}}{\text{assimilation}}$, is commonly used by those interested in the amount of potential food that a population can produce. This is the efficiency with which energy is fixed in the organic matter of a population or trophic level and made potentially available as food to other populations or trophic levels.

The ratio of respiration to assimilation is also used. To the energy assimilated in a time unit must be added the energy equivalent of any decrease in standing crop within the period since such a decrease represents a mobilization of energy previously assimilated and stored in the bodies of organisms.

The ratio of respiration to assimilation is also used. R. S. Miller for their interest and helpful suggestions, also to Selwyn Roback, H. K. Townes, C. J. Goodnight and Arthur Clarke for help in identifying the fauna. The cooperation of the Root family who own the spring is gratefully acknowleded.

The Cold Spring

Small, constant temperature springs are as nearly perfect systems for the study of community metabolism as can be found in nature. They have the advantage of a comparatively unchanging chemical and physical environment, which reduces the difficulties of measurement and makes laboratory experiments simpler, for it is easier to duplicate constant conditions than varying ones. Also, the biota in cold springs has fewer species than do most communities.

In spite of these advantages, the ecology of springs, especially cold springs, has not received much attention in the United States. The faunas of the hot springs of this country were studied by Brues (1928). Dudley (1953) has investigated the faunas of some springs of varying temperature and Odum has worked with the rich cold springs of Florida (Odum 1957).

The spring chosen for this study of community metabolism is a limnocrene, a spring in which the water emerges into a basin, located on Intervale Farm belonging to the Root family in Concord, Massachusetts. It lies at the foot of a bank of glacial till which extends laterally for about one mile and from which emerge a number of rheocrenes, springs which form brooks immediately. The basin of Root

Spring is about 2 m in diameter and the water, which comes out of the ground around the uphill edge and flows out in a springbrook on the opposite side, is 10-20 cm deep. Most of the bottom is covered with mud and it is in the mud that the organisms are found. As is the case with most springs there is no true plankton.

The spring was sampled from June 1953 until November 1954, although general observations extended from February 1953 to March 1955.

November 1953 through October 1954 was the period chosen for the analysis because the emergence of the insects was over for the year by November and hence egg laying by the insects was also completed.

Environmental Conditions.—Figure 1 gives a summary of the conditions within the spring from August 1953 to July 1954. Since there was no very unusual weather during this year, it seems likely that the observed conditions were typical. The fauna and flora were subject to little in the way of changes in their physical and chemical environment. The temperature varied at most 2° C from the mean annual air temperature for the Concord region, 9.5° C. The high concentration of CO_2, 20-30 ppm, had no adverse effect on the spring fauna as far as could be observed although it may have had an important effect in excluding intolerant species. The same may be said for the oxygen concentration which ranged from 26-65% of saturation.

The Flora.—The flora of Root Spring from June 1953 to November 1954 consisted entirely of benthic algae and the duckweed, Lemna. During November and December there were only a very few diatoms present. In January the flora began to increase and the first species to appear in abundance was one of the diatoms, Eunotia, which grew on all of the available solid surfaces. As the amount of light per day increased, filamentous green algae appeared in masses all over the bottom of the spring. *Stigeoclonium stagnatile* was the principal species, along with a smaller amount of Spirogyra sp. By May these species had decreased in numbers and a colonial green alga, *Tetraspora lubrica,* and a tiny diatom, *Nitzschia denticula* (?), made up the main biomass of the flora. As summer progressed the green algae all diminished in abundance and the benthic diatoms were the only plants of importance. In the autumn Spirogyra, and Oedogonium and *Coleochaete soluta* formed a considerable part of the green plant flora along with the diatoms.

Lemna covered the edges of the water out of the current in the early part of the year and was mostly gone by April. It did not seem to contribute much to

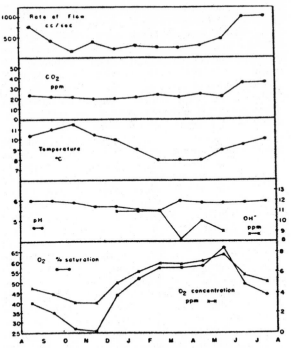

FIG. 1. Physical and chemical characteristics of the water in Root Spring, Concord, Mass., 1953-54. All data are plotted as monthly averages.

the spring community because, instead of sinking to the bottom of the springpool, it was washed out of the outlet and down the springbrook when the rains of March and April agitated the water.

While the algae contributed a considerable amount of energy to the animal populations, the main source of food, as will be shown below, came in the form of plant debris, mostly leaves from apple trees, which collected on the bottom of the spring.

The Fauna.—Although over 40 species of animals were identified from Root Spring, there were relatively few species that occurred in numbers and sizes large enough to be important in the energy balance. The most abundant animals were those which fed on debris and algae, taking mud into their gut and assimilating the digestible material. These included the oligochaete, *Limnodrilus hoffmeisteri;* and the chironomid larva, *Calopsectra dives.* Also feeding upon detritus and debris were the isopod, *Asellus militaris,* the amphipod, *Crangonyx gracilis,* and the fingernail clam, *Pisidium virginicum.* The snail, Physa, feeds on detritus and on algae which it scrapes off of the surface of the mud. The caddis larvae,

Frenesia difficilis, F. missa, and Limnophilis sp. eat larger bits of vegetation.

Another chironomid, *Anatopynia dyari,* eats other animals but when this food is in short supply it can get along on plant material. Other predators were the planarians, *Phagocata gracilis* and *P. morgani.*

Root Spring differed from the characteristic cold spring described by Pennak (1954) in lacking leeches and black fly larvae and having planarians in abundance.

A list of all of the organisms found in the spring which were identified follows:

Algae
  Chlorophyta
    Stigeoclonium stagnatile (Hazen)
    Coleochaete soluta (Brebisson)
    Tetraspora lubrica (Roth)
    Spirogyra sp.
    Oedogonium sp.
    Closterium sp.
  Chrysophyta
    Cymbella aspera (Ehrenberg)
    Nitzschia denticula Grunow
    Stauroneis phoenicentron (Nitzsch)
    Eunotia sp.
  Cyanophyta
Higher plants
    Lemna minor L. (?)
Protozoa
    Euplotes sp.
    Paramecium sp.
    Rhabdostyla sp.
    Vorticella sp.
Platyhelminthes
    Phagocata gracilis woodworthi Hyman
    Phagocata morgani (Stevens and Boring)
Gastrotricha
    Chaetonotus sp.
Rotatoria
    Trichocerca sp.
    Lepadella sp.
Nematoda
    Monhystera sp.
    Plectus sp. (?)
Nematomorpha
    Gordius sp.
Annelida
  Naididae
    Chaetogaster langi Bretscher
  Tubificidae
    Limnodrilus hoffmeisteri Claraparede
    Peloscolex sp.
Mollusca

  Pisidium virginicum (Gmelin)
  Physa sp.
 Crustacea
  Copepoda
   Cyclops vernalis Fischer
   Canthocamptus staphylinoides **Pearse**
   Eucyclops agilis (Kock)
   Paracyclops fimbriatus (Fischer)
  Ostracoda
   Eucypris sp.
  Isopoda
   Asellus militaris Hay
  Amphipoda
   Crangonyx gracilis Smith
 Arachnoidea
 Hydracarina
  (Parasite of *C. dives*)
Insecta
 Collembola
  Tomocerus sp.
 Ephemeroptera
 Megaloptera
  Sialis sp.
 Trichoptera
  Frenesia difficilis (Milne)
  Frenesia missa (Walker)
  Limnophilis sp.
  Lepidostoma sp.
 Coleoptera
  Agabus sp.
  Bidessus sp.
 Diptera
  Calopsectra dives (Johannsen)
  Calopsectra xantha Roback
  Anatopynia dyari (Coquillett)
  Anatopynia brunnea Roback
  Pentaneura carnea (Fabr.)
  Prodiamesia olivacea (Meigen)
  Tendipes tuxis Curran
  Brillia parva Johannsen
  Metriocnemus hamatus (Johannsen)
  Hydrobaenus obumbratus (Johannsen)
  Tanytarsus fuscicornis (Malloch)
  Erioptera septemtrionis O.S.
  Psychoda sp.
  Eucorethra sp. (?)
  Pericoma sp.
  Culex apicalis Adams
  Mansonia perturbans (Walker)
 Hymenoptera
  (A parasite of *C. dives*)

METHODS

Population Estimates.—A Dendy sampler (Welch 1948) was used to obtain samples of the bottom animals. A map of the spring was covered with a numbered grid and successive sampling sites were chosen from a table of random numbers. The samples were placed in a 16 mesh per centimeter sieve and the mud and fine debris removed by washing in the spring.

The animals were separated by hand under a magnifying lens from the mass of tubes and debris that remained. The individuals of each species were counted and the live weight determined after removal of excess water with filter paper.

The Phagocata presented a special problem because they secreted mucus when handled and because their epidermis permitted rapid water loss. They were placed on a fine screen and excess water was removed with filter paper. The screen with the animals was then weighed, the animals quickly removed with forceps, and the screen reweighed.

After counting and weighing the animals, with the exception of less than 1% used in experiments, were returned to the spring in order to minimize disturbance of the community.

Weight was converted to calories by oxidation with potassium dichromate in sulfuric acid (Ivlev 1934). The nitrogen content of the animals, needed for this method, was taken from the literature or found by Nesslerization.

The numbers of adult insects that emerged from the water from April to October were measured with tent traps of $\frac{1}{3}$ m^2 or $\frac{1}{10}$ m^2 area.

Some attempt was made to determine the relative number of bacteria in the spring. This was done following the method of Henrici (1936) by putting clean slides about one-fourth of their length into the mud and determining the time required for a coating of bacteria to grow.

Respiratory Measurements.—Rates of respiration of the principal animals were measured by the method of Ewer (1941), with the spring serving as a constant temperature bath. Animals were taken from the spring with as little disturbance as possible and quickly transferred to 20 cc syringes filled with spring water. The oxygen content of the water was then measured with the Micro-Winkler technique (Fox & Wingfield 1939). The syringes were placed in the spring and after sufficient time had elapsed for a measurable change to occur but not so long that oxygen tension was appreciably lowered (1-3 hrs) the oxygen content of the water was again determined. During the interval the water was usually kept sufficiently stirred by the activities of the animals themselves but if this was not the case, the syringes were turned over at regular intervals. This procedure measured the respiration under the same conditions of temperature, oxygen, pH, alkalinity, etc. as those which the animals normally experienced. At the end of the measurement the animals were taken to the laboratory and weighed. Oxygen consumption was converted to calories with the average oxycaloric coefficient of Ivlev (1934), 3.38 calories per milligram of oxygen. This average coefficient was used as the respiratory quotient was not known.

Since two of the species studied, Calopsectra and Limnodrilus, normally live in tubes, measurement of their oxygen consumption with the animals out of their tubes and in the syringes could be subject to error. Walshe-Maetz (1953) found, however, that while the oxygen consumption of *Chironomus plumosus* was different at oxygen concentrations below 25% saturation if the animals were removed from their tubes, at higher concentrations there was no significant difference. Since the oxygen concentration in the spring never fell below 25% saturation, the above possibility of error may be safely neglected.

Molting Losses.—When an arthropod molts it leaves a certain amount of energy behind in the material of the cast off exoskeleton. Midge larvae were kept in the laboratory and weighed daily. The loss of weight after a larva-to-larva molt represents the amount of material lost with the cast skeleton. The larva-to-pupa-to-adult transition was treated as one process. The loss of weight between the prepupal larva and the adult represents the material lost in the larva-to-pupa molt, the material lost in the pupa-to-adult molt and loss due to pupal respiration.

Oxygen Changes Due to Respiration and Photosynthesis.—Changes in oxygen due to the combined activities of all organisms living in the spring were measured by determining the changes in oxygen in the water enclosed by glass cylinders 17 cm in diameter which were pushed into the mud until they projected only 2-4 cm above the water surface. They were then filled to the top with water and covered with a glass plate. The top edge of the cylinder was ground so that the fit would be air tight. The water was sampled from all levels within the cylinder and the oxygen content measured by the standard Winkler technique at the beginning and again after a period of about 24 hours. Oxygen gradients that may have formed within the cylinder caused no appreciable sampling error. Their possible effect on the organisms was ignored.

Respiration was then measured during a period

of several hours with a black box covering the cylinder to prevent any photosynthesis. By subtracting the respiration due to the macrofauna, the respiration of the decomposers, microfauna, and algae was determined.

Portion of Food Assimilated.—To find out how much potential food was not assimilated in the feeding process, Anatopynia larvae, kept at 9° C in spring water, were weighed and given weighed amounts of live Calopsectra or Limnodrilus every second or third day. The respiration of Anatopynia during the experiment was measured.

The portion of food assimilated was calculated from these measurements on the assumption that the difference between the energy contained in the food and the energy contained in the tissue added by growth plus the animal's respiration represented loss on the form of feces and in the form of material from the prey which was killed but which never entered the predator's gut.

The feeding efficiency of Phagocata was determined in a similar manner. There was a difference, however, in that most of the loss which occurred during the feeding activities of Phagocata was due to the large quantities of mucus that were secreted by the animals in their movements in search of food and in their actions in subduing their prey.

Anatopynia larvae are stated to be carnivorous (Johannsen 1937, Pennak 1954) but the present observations showed that they also digested plant food. Larvae were anesthetized each month, the contents of their guts examined under a compound microscope and the portion of animal and plant food estimated. As much could be seen in the gut of the live animal as could be observed by examining the guts of intact, cleared animals or examining food removed from larval guts. Only a relative measure of the amounts of the various types of food that were taken could be obtained with any of these methods. Difficulties were encountered in identifying the contents of the gut and in estimating the amount of food that was represented by the portions that remained.

Measurement of Mortality.—In order to determine completely the energy flow through a population it is necessary to know what quantity of energy is lost by the death of organisms from whatever cause between periods of sampling.

Each month's sample mean was assumed to represent the actual population on the midday in the month. A sampling interval extended, therefore, from the 16th of one month to the 15th of the next.

An estimate of the mortality between samples was obtained with the use of the equation given by

Ricker (1946): $P_t = P_o e^{(k-i)t}$ in which k is rate of growth or net production; i is rate of mortality; P_o is weight of population at time zero; P_t is weight of population at time t; t is taken to equal one month.

Values for k were determined by raising animals in a 9° C cold room and by estimation from the greatest rate of increase that occurred in the natural population. In those populations which did not reproduce during the summer, k was determined from the average increase in weight of surviving individuals.

Once k was determined it was used as a constant (except in the cases of Calopsectra and the Trichoptera where it was determined each month by the last method mentioned above). Net production was probably about the same at all seasons of the year for most animals in the constant conditions of the spring. It makes no difference in calculating energy flow whether the net production of a population takes the form of growth of individual organisms or of increase in numbers of individuals (i.e., reproduction).

A value for i was found for each sampling interval from the above equation and multiplied by the average population (geometric mean) during the interval to give an estimate of mortality. This mean is not so exact as that given by Ricker (if i is constant for the interval considered) but it is easier to calculate. Actually, i is a continuous varying quantity rather than constant for any month and use of the geometric mean is accurate enough for this study.

Since the data for energy flow were compiled on basis of the calendar months, the total mortality was divided between the two months involved according to the ratio of the geometric means of the populations present in the second half of the first month and the first half of the second month (assuming each month to contain 30 days). The mortality is a minimum estimate since the mean of the populations present during each was used in the calculations as if it were the true value on the fifteenth of the month. This use tends to smooth out the curve of population size.

Input of Energy in the Form of Organic Debris.—
The amount of debris that fell into the spring was estimated by placing a box having sides 30 cm high and a bottom area of $\frac{1}{10}$ m² on the ground next to the spring and collecting the material that accumulated in the box. This material, mostly leaves, twigs, fruit, etc., was dried and weighed and its caloric content determined.

The organic matter in the inflow and outflow of the spring was measured following the method of Pennak (1946).

SPECIES POPULATIONS

The results will be presented and discussed separately for each species that was individually studied. These data will then be combined to give a picture of the community metabolism for the spring as a unit. The actual sampling data may be found in a PhD thesis deposited in the Biology Library, Harvard University.

Calopsectra (= *Tanytarsus*) *dives* (Johannsen).—This species is one of the two abundant chironomids in the spring and is a typical member of this genus characteristic of oligotrophic waters. The population of Calopsectra during the study is shown in Fig. 2 and Table 1. There were no larvae of this species during the winter. A maximum standing crop of 87.6 KC/m^2 was reached in August after which pupation and emergence of adults reduced the population again to zero. The population means for June, July and August differ significantly at $P = 0.01$. The calories contained in the standing crop were found to be 0.69 KC/gm $\pm$ 0.020 (12 determinations).

Fig. 2 also shows the number of adults that emerged and the monthly totals are given in Table 1. The sex ratio among the adults was 53% female and 47% male (based on 1500 adults).

Because mating and egg laying take place so soon after emergence it was estimated by observing the insects that about two-thirds of the females laid their eggs in the spring from which they had just emerged and the others dispersed. By counting the eggs contained in 5 virgin females it was found that they laid an average of 250 eggs each.

The number of eggs that gave rise to the 1954 population was calculated from emergence data for

TABLE 1. Population of larvae and emerging adults of *Calopsectra dives* in Root Spring, Concord, Mass., in 1954. Energy content is estimated on the basis of 1.58 cal/mg as determined for *Anatopynia dyari*.

	Jan.-April	May	June	July	Aug.	Sept.	Oct.-Nov.
Larvae							
Number in thousands/m^2	..	1.7	89.5	65.0	57.0	0.2	......
Weight in gms/m^2	..	3.0	58.5	82.4	127.0	0.2	......
Energy content in KC/m^2	..	2.1	40.4	56.8	87.6	0.1	......
Adults							
Number/m^2	..	13	170	953	3464	13250	533
Weight in mg/m^2	..	12	156	876	3170	12200	490
Energy content in KC/m^2	..	0.019	0.246	1.38	5.00	19.30	0.775

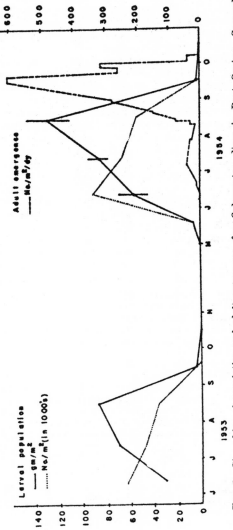

FIG. 2. Size of larval population and adult emergence for *Calopsectra dives* in Root Spring, Concord, Mass. Vertical lines represent 95% confidence limits.

1953. Fifty-eight hundred females emerged at that time and these females at 250 eggs apiece laid roughly 980,000 eggs per square meter.

The mortality of larvae calculated from Table 1 for July to August was 15% and for June to July, 28%. Thirty percent, therefore, will be used for mortality from May to June when the larvae were smallest. This figure agrees with the mortality rates of young chironomid larvae given by Borutzky (1939). Most of these larvae did not appear in the samples in May because they were so small that they passed through the meshes of the sieves. On the basis of 30% larval mortality 128,000 larvae emerged from eggs in May. As 980,000 eggs were laid, 87% of these did not survive to hatch.

As no new larvae hatched during the summer, all of the individuals are of about the same age and mortality is easily calculated with the formula:

$$\text{Mortality} = \text{antilog} \frac{(\log P_o - \log P_t)}{2} \frac{(\log P_p - \log P_t)}{\log e}.$$

The first expression gives the average population present during the interval and the second gives the death rate. P_o is the size of the population at the beginning of the instar; P_t the size at the end of the instar; and P_p the theoretical size of the population at the end of the instar had there been no mortality. To facilitate the calculation, Table 2 has been set up.

TABLE 2. Data for calculation of mortality of *C. dives* larvae (per square meter). The value of P_o for May is taken from the total estimated larvae hatching multiplied by the calories per egg (.002). The value for P_t in August is the total calories contained in the larvae that successfully emerged as adults.

	Instar	No. of larvae	P_o	Gram cal. larvae at end instar	P_t	P_p
May 15 to Jan 15...	1	128,000	0.256KC	.45	40.4KC	57.6KC
Jun 15 to Jul 15....	2	98,500	40.4	.86	56.8	76.5
Jul 15 to Aug 15...	3	65,000	56.8	1.54	87.6	103.1
Aug 15 to Sep 15...	4	57,000	87.6	2.09	28.8	119.1

Mortality was divided among the calendar months as described previously except for that calculated for the last instar which was divided among August, September and October in proportion to the adult emergence. Most of the deaths during this interval occurred as the pupae tried to wriggle out of the larval tubes and were caught in the mucus which the increased numbers of planarians had spread over the tube mouths.

The energy lost by respiration is, in a sense, the most important part of the energy flow through the population since this is the energy that the animals actually use in their life processes. From 9 measurements the respiratory rate of Calopsectra was found to be 0.475 ± 0.022 mg O_2/gm/hr or 1.67 ± 0.08 cal/calorie of larva per month.

With the data presented above it is possible to draw up an energy balance sheet for the population of *Calopsectra dives* during the summer of 1954 (Table 3). Calculation of the respiration of animals that died between samplings was based on an average life of one-half month (i.e. mortality uniform with time). The sum of the energy assimilated by the population of *Calopsectra dives* was 520.3 kilo-calories per square meter. The larvae transformed 389.6 KC, about 75% of the input.

Some of the energy carried out of the system in the bodies of the adults was used in the life processes of the adults and some of the energy contained in the female adults was in the eggs and would have been used by the developing embryos. Therefore, about 80% of the input energy was transformed to heat by the midges.

The efficiency, $\frac{\text{net production}}{\text{assimilation}}$, of the Calopsectra was 100% minus 80% = 20% since this is the amount of energy that the population passes on to other populations.

Anatopynia dyari (Coquillett).—Anatopynia is the second most abundant species of midge occurring in the spring. Figs. 3 and 4 and Table 4 show the size of the population of Anatopynia larvae which varied from a low of 4.4 KC/m^2 in September during pupation and adult emergence to a high of 28.2 KC/m^2 in October. These figures include two species other than *A. dyari; Anatopynia brunnea* Roback and *Pentaneura carnea* (Fabr.). Both of these latter species are similar in habits to *A. dyari* and were less than ⅕ as numerous.

The energy content of the larvae was found from 6 determinations to be 0.88 ± 0.056 KC/gm.

The differences in numbers of individuals between May and June, June and July, July and September 1954 and October and November 1953 were significant at $P = 0.01$ and between September and October 1954 at $P = 0.05$.

It was estimated from field observations that about ¼ of the emerging females lay their eggs in the spring. Since there are no other springs in the vicinity from which Anatopynia adults could reach Root Spring, 264 females which emerged in May and June deposited 23,100 eggs and 980 females which emerged in August and September deposited 86,100 eggs. (Five virgin females were examined and found to contain 325, 340, 350, 370, and 375 eggs, or an average of 350 eggs each.)

By examination of the graph of the numbers of individuals of *A. dyari* (Fig. 3) the mortality of the eggs and the very young larvae can be estimated. From November through May 15 there were about 7500 larvae/m^2. From May 15 through July, 510 adults emerged but between May and June there was a decrease in the population of 4500 larvae/m^2. This indicates that 3990 larvae died in this interval. This decrease is reflected in the curve for weight/m^2. Newly hatched larvae did not appear in the samples until about one month after the eggs were laid.

In July the numbers/m^2 increased to 5000, representing an addition of 2000 new larvae from the 23,100 eggs laid during May and June. A mortality of 91.4% occurred among the eggs and young larvae. In November somewhat more than 10,000 new larvae appeared indicating a mortality of 88% of the autumn eggs and young larvae. On the basis of a three to one ratio for the mortality of eggs to the mortality of young larvae (from the data of Borutzky 1939

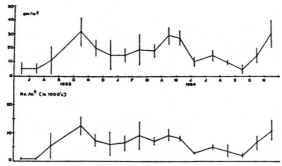

Fig. 3. Population of Anatopynia larvae in Root Spring, Concord, Mass. (The vertical line indicates the 90% confidence limits of the mean.)

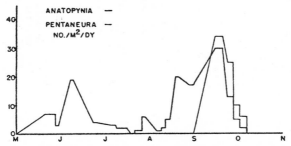

Fig. 4. Emergence of adults of Anatopynia (heavy line) and Pentaneura (light line) from Root Spring, Concord, Mass. in 1953-54.

TABLE 3. Balance sheet for energy flow in *Calopsectra dives* in Root Spring, Concord, Mass., in kilo-calories per square meter. Column 4 shows the calories respired before molting by the biomass represented in the deposited larval skins. Values for deposits were measured for *A. dyari* and assumed proportional for *C. dives*.

Month	Standing crop	Respiration of S. C.	Respiration of animals that died	Respiration of deposit	Emergence	Pupal deposit	Larval deposit	Mort.
October to April	...	...	...	...	...	...	...	...
May	2.1	3.5	...	...	0.02	0.01	...	...
June	40.4	67.5	0.6	3.3	0.25	0.11	4.0	0.7
July	56.8	95.0	1.0	4.8	1.38	0.61	5.7	1.2
August	87.6	146.2	12.4	7.3	5.00	2.2	8.8	14.9
September	0.1	0.2	45.8	...	19.30	8.5	...	54.8
October	...	...	1.8	...	0.78	0.34	...	2.1
		312.6	61.6	15.4	26.7	11.8	18.5	73.7

Net change in standing crop....0
Total respiration............389.6
Total assimilation...........520.3

TABLE 4. Monthly summary of larval population and adult emergence of *Anatopynia dyari* and *Pentaneura carnea* in Root Spring, Concord, Mass, 1953-54. Larvae of the two species were not distinguished.

	1953			1954										
	Oct.	Nov.	Dec.	Jan.	Feb.	Mar.	Apr.	May	Jun.	Jul.	Aug.	Sep.	Oct.	Nov.
Larvae														
Number in 1000's	12.4	7.2	6.0	6.3	8.9	6.8	9.1	8.1	2.6	5.1	3.8	2.0	6.6	10.9
Energy content in KC/m²	28.2	17.6	13.2	12.3	15.8	15.8	25.5	16.7	9.7	13.2	8.8	4.4	13.2	27.3
Adult														
Anatopynia														
No/m²	...	...	...	...	...	...	...	128	308	73	308	633	15	...
Mg/m²	...	...	...	...	...	...	...	311	599	129	642	1188	31	...
Pentaneura														
No/m²	...	...	...	...	...	...	...	...	...	...	...	646	40	...
Mg/m²	...	...	...	...	...	...	...	...	...	...	...	667	41	...

for another species of chironomid), it was calculated that these larvae stored a total of about 0.4 KC/m^2 in their bodies before they died, and this amount is included in the tabulation of energy flow. This calculation is very rough since exact data are not available, but it is certain that the effect of these young larvae did not materially affect the total energy flow for this species, 239.2 KC/m^2.

A few determinations were made of the amount of energy lost in the molted exoskeletons of the larvae. *A. dyari* was used for these measurements because of their large size and good viability under repeated handling. Three larva-to-larva molts were successfully measured. The loss was 11.5%, 10% and 9%, or an average of 10% loss of energy/molt in terms of the final larval weight. This agrees well with the figure of Borutsky (1939) for *Chironomus plumosus* who found a loss of 16.8% in all three molts in terms of the final larval weight. A loss of 10%/molt is equal to 16.3% loss in three molts.

There is a similar loss of energy in the transformation from prepupal larva to pupa and then to adult. It was found that an adult midge came from a larva which contained 1.44 times as many calories as the adult. Thus 0.44 cal/cal of adult were deposited with each adult emergence.

It was apparent, after examination of larvae from the spring and raising larvae successfully on a diet of Spirogyra, that *A. dyari* larvae are both herbivores and carnivores. The larvae may not, however, be able to exist solely upon detritus. During December and January when algae were scarce and the larvae were not obtaining much animal food, their average weight decreased. The approximate portion of the food of *A. dyari* larvae which came from animal sources was as follows: none in January; one-eighth in April, one-quarter in November, December, March, July and October; one half in May, June and August; three-quarters in February and September.

It will be seen when the predator-prey balance for the entire community has been presented that the estimation of the proportions of foods from the two sources must be at least approximately correct.

When detritus-feeding forms eat, they take into their guts a variety of material usually in small bits and their digestive enzymes act on the mass of material to make available for absorption all the potential food. The only organic material that passes through is that which is indigestible, considered by many workers to be the crude fiber portion of the organic matter (Lindeman 1942, Birch & Clark 1953). When the food is animal prey, however, unless the predator can swallow the prey whole, a certain amount of it will be lost without ever entering the alimentary canal.

This waste was measured for *A. dyari* feeding on Calopsectra and Limnodrilus.

Thirty duplicate experiments revealed that when feeding on animals *A. dyari* are able to assimilate, on the average, only 30% of their prey (29% ± 2%). The remaining 70% is lost in the form of (1) blood of the prey which passes into the water, (2) flesh which is not swallowed by the predator, and (3) indigestible material which passes through the predator in the form of feces. This figure seems reasonable when it is realized that Anatopynia grab their prey in their mandibles and draw it into their esophagus bit by bit. The skin of the prey is ruptured in the process and blood escapes. Since the mouth of Anatopynia is smaller than the usual prey, a large amount of the flesh is also lost. The loss may be less when Anatopynia feeds on animals which are small enough to be swallowed whole.

Measurements of respiration of the larvae of this species were made in Root Spring throughout the year. Eighteen measurements gave a mean of 0.45 ± 0.04 mg O_2/gm fresh wt/hr or 1.24 calories respired/calorie of larvae/month. The figures for respiration in the balance sheet (Table 5) are corrected for mortality and deposit as explained for *C. dives*. The net change in the standing crop for the year was found by subtracting from the increases the decreases due to deposit, emigration, and mortality. Mortality was calculated from the differences in the standing crop and deposit and emigration. Mortality and increases in the standing crop were divided between months in proportion to their standing crops. Energy obtained from cannibalism was figured as ⅓ mortality on the basis of the feeding efficiency. Except for August when there were a large number Phagocata present, the mortality of Anatopynia was probably due to cannibalism since this chironomid was the most abundant predator present. In December and January the mortality may only be respiratory loss as there was no significant decrease in numbers during those months.

The total assimilation of energy by Anatopynia larvae for the year of the study, 249.1 KC/m², was found by adding losses due to respiration, death, deposits of larval and pupal exuvia, and emigration of adults. Of this total assimilation 9.9 KC came from cannibalism and 3.2 KC from a decrease in the standing crop, leaving 236.0 KC which came from sources outside the Anatopynia population.

The proportion of energy transformed by Anatopynia is $\frac{208.5 \text{ KC respired}}{249.1 \text{ KC assimilated}}$ or 84%. Their efficiency at passing energy containing material to other popu-

TABLE 5. Energy balance sheet for Anatopynia in Root Spring, Concord, Mass. for 1953-54 (in kilo-calories per square meter).

Month	Standing crop	Larval deposit	Pupal deposit	Emigration	Mortality	Energy from Cannib.	Respiration	Increase in S. C.	Energy from animals	Energy from algae
Oct. '53	(28.2)	(1.4)	...	...	...	...	...	...	...	...
Nov.	17.6	1.0	...	...	6.1	2.0	21.4	...	4.9	14.5
Dec.	13.2	...	...	...	2.5	2.5	16.4	...	3.5	10.4
Jan. '54	12.3	...	...	...	0.4	0.4	15.3	...	...	16.5
Feb.	15.8	...	...	...	...	...	19.6	1.6	...	5.4
Mar.	15.8	...	...	...	...	...	19.6	1.9	16.1	17.9
Apr.	25.5	1.6	0.2	0.5	4.5	1.5	32.6	4.3	6.0	33.3
May	16.7	1.0	0.4	1.0	6.9	2.3	21.3	8.0	4.8	9.5
Jun.	9.7	...	0.1	0.2	1.3	0.4	12.0	...	9.5	6.7
Jul.	13.2	...	...	1.0	2.6	0.8	16.6	1.7	6.6	13.6
Aug.	8.8	...	0.4	...	3.1	(1.0)	10.8	2.3	4.5	5.4
Sep.	4.4	...	1.3	2.9	...	...	5.5	5.5	5.4	2.8
Oct.	13.2	1.3	...	0.1	0.2	...	17.4	13.1	8.2	22.9
Nov.	(27.3)	...	...	...	...	...	...	...	7.6	...
	−3.2	4.9	2.4	5.7	27.6	9.9	208.5	37.4	77.1	158.9

105

lations is $\dfrac{30.7\text{ KC passed on to other populations}}{236.0\text{ KC assimilated from outside sources}}$ or 13%.

Since Anatopynia larvae are both herbivores and carnivores, the energy flow through them may be divided into two parts in proportion to the energy obtained from animal and plant sources. This was done in the balances for these two trophic levels presented below.

Limnodrilus hoffmeisteri Claraparede.—Limnodrilus is a typical aquatic, tube dwelling oligochaete, which reproduces throughout the year in Root Spring. It feeds upon detritus.

Table 6 gives the monthly population of Limnodrilus. The caloric content was found to be 0.76 ± 0.026 cal/mg fresh wt. The great decrease in population in November 1954, just after the study period, was probably caused by the action of the several hurricanes of that fall, which disturbed the spring by uprooting nearby trees.

The amount of energy transformed by Limnodrilus, measured by respiration, was calculated to be 1.22 ± 0.35 cal/cal fresh wt/month (0.83 ± 0.024 mg O_2/mg/hr) from 15 determinations.

The mortality was calculated using the formula of Ricker (1946) by making the assumption that the rate of increase was constant throughout the year. Since the temperature of the water is nearly constant the year round and since it makes no difference for this calculation whether the increase takes the form of reproduction or growth, this assumption of constant rate of increase is probably quite reliable. It may be noted that the worms had full guts at all seasons.

To find the rate of increase, k, the two methods mentioned previously were used. Weighed animals confined in strained mud in the laboratory cold room were reweighed after two weeks. When one or more animals in an experiment died, the experiment was discarded. Of the 5 successful experiments the one with the maximum rate of increase, in which k equaled 0.474 for a 30 day month, was taken as the significant one since laboratory conditions were not as conducive to growth as were conditions in the spring to which this population of worms was adapted.

A value for the rate of increase was also obtained from the increase of the natural population in the spring from April to May, $k = 0.530$. Even though the P value lay between 0.05 and 0.10 for this population increase, the larger value for k was used as it agreed fairly well with the maximum value obtained in the laboratory experiments and as k obtained

TABLE 6. Population sizes in numbers per square meter and kilogram calories per square meter for animals other than chironomids in Root Spring, Concord, Mass. in 1953-54.

	1953				1954									
	Oct.	Nov.	Dec.	Jan.	Feb.	Mar.	Apr.	May	Jun.	Jul.	Aug.	Sep.	Oct.	Nov.
Limnodrilus														
No. of animals	?	7600	7700	9000	10600	4800	4200	7300	4300	3700	3900	3450	3200	1300
Energy	23.0	32.0	30.4	32.0	53.0	17.6	18.0	30.6	16.7	16.4	23.2	21.3	18.7	3.7
Asellus														
No. of animals	1700	270	500	1100	400	100	400	100	100	200	2400	1700	900	1200
Energy	16.6	4.3	8.6	26.5	8.4	1.8	4.9	1.9	2.3	2.8	11.4	10.0	8.1	19.4
Phagocata														
No. of animals	1000	200	500	3300	1800	1100	300	400	500	900	2400	4600	1500	100
Energy	7.1	2.3	2.7	22.4	5.1	3.9	2.4	0.8	1.3	3.1	16.2	27.9	6.8	1.2
Trichoptera														
No. of animals	...	...	...	980	390	270	470	710	350	80	200	...	...	...
Energy	...	...	...	0.78	0.78	1.14	3.41	9.73	7.45	2.39	10.8	...	...	...
Pisidium														
No. of animals	1000	500	470	1300	2700	1500	2700	1500	1100	1900	1400	2400	3600	800
Energy	4.0	1.7	2.0	6.8	6.5	4.8	9.3	5.5	6.4	6.6	4.7	5.7	6.8	3.9
Physa														
Energy	0.2	0.8	1.9	3.6	1.4	1.1	1.1	0.7	1.3	...	4.1	...	0.1	...
Crangonyx														
Energy	...	...	0.2	1.4	0.9	0.6	1.2	0.7	0.8	0.3	4.1	0.7	...	...

TABLE 7. Energy flow figures for the animals of Root Spring, Concord, Mass., 1953-54, with the exception of the chironomids. All data in kilocalories per square meter per year.

Species	Change in S. C.	Respiration	Immigration	Mortality	Mucus loss	Cannibalism	Outside assimilation	Total energy flow	Net Production
Limnodrilus	−13.3	483.6	...	173.8	...	...	644.1	657.4	173.8
Asellus	3.8	486.1	...	104.5	...	...	604.4	604.4	104.5
Phagocata	−0.9	18.7	...	48.2	89.1	23.9	131.2	156.0	113.4
Trichoptera	...	67.5	18.3	39.2	?	...	88.4	106.7	39.2
Pisidium	5.1	90.9	...	76.7	...	...	172.7	172.7	81.8
Physa and Crangonyx	...	90	...	30	...	...	120	120	30

from the fluctuations of a natural population will have a minimum value. The balance sheet for the energy flow through the population of Limnodrilus was set up in Table 7.

These tube worms assimilated a total of 644.1 KC/m^2 during the year of investigation. Some energy assimilated previously also flowed through the population since the standing crop decreased by 13.3 KC/m^2.

The oligochaetes used, in their life processes, $\frac{173.8\ KC}{657.4\ KC}$ or 74% of the energy assimilated. They passed on to other populations 173.8 KC, an efficiency of $\frac{173.8\ KC}{657.4\ KC}$ or 26%.

Asellus militaris Hay.—*A. militaris* is the most common and widespread of the American species of aquatic isopods. It feeds on anything edible that it encounters although it does not ordinarily kill prey. In the Root Spring the isopods reproduced throughout the year. The population size is given in Table 6.

The figure for the conversion of moist weight to calories was calculated from the analysis of the chemical composition of Asellus by Ivlev (1934).

Mortality was figured using a value for k calculated from the increase in the population from July to August. The difference between these two means is significant at $P = 0.05$. An error could arise from migration into the spring since this is one of the species that could crawl up the springbrook from the pond below. However, no Asellus were even found in the springbrook and migration was probably insignificant. Losses due to respiration were found to be 1.22 ± 0.05 mg O_2/mg/hr.

There is an error in the energy balance due to the fact that no allowance was made for energy lost by molting in this crustacean. The error would not come to more than 10% of the energy passed through the population and would probably be less to judge from the molting of the chironomids for which this calculation was made.

The energy balance for this species is presented in Table 7. Some energy, 3.8 KC, was stored in an increase in the standing crop during the year of study. About 80% of the energy intake of Asellus was used for its life processes.

Phagocata gracilis woodworthi Hyman and *P. morgani* (Stevens and Boring).—These two planarians are the only important exclusively predatory animals in the spring and feed on live or recently dead animals. Since they are able to suck in only the softer parts of their prey, the harder parts, such as the exoskeletons of arthropods, are left behind. These species were observed to have no definite breeding season and reproduced throughout the year.

Both species of Phagocata are considered together in the energy flow calculations because they are ecologically similar. They feed on the same sorts of material and live in the same habitat. The fact that they are members of the same genus also indicates probable similarity in ecology. They differ in size, however; *P. morgani* is seldom found to weigh more than 1-2 mg while *P. gracilis* attains a weight of 20 mg. They also differ in that *P. gracilis* has many pharynxes while *P. morgani* has only one, and in that most of the apparent "cannibalism" among flatworms in the spring is really due to *P. morgani* feeding on *P. gracilis*. The size of the combined populations is given in Table 6.

The calories per unit weight were calculated from data gathered for *P. gracilis* and assumed to be the same for *P. morgani* (1.33 ± 0.02 cal/mg). The rate of respiration was also calculated using *P. gracilis* as the experimental animal and was 0.0735 ± 0.005 mg O_2/gm fresh wt/hr. Starvation of planarians causes their respiratory rate to vary (Hyman 1919) but was not a factor in these determinations since the animals were taken at random from the natural population.

Flatworms differ from other sorts of animals in that they secrete a great deal of mucus in their activities. They lay down a film of mucus whenever they move over objects and use it to ensnare their prey. For this reason it was necessary to measure the amount of energy lost by the Phagocata in the form of mucus.

Food in the form of weighed amounts of live oligochaetes was given to the worms at intervals far enough apart so that as far as could be determined by observation all of the food was consumed. It was then assumed that the difference between (1) food supply and (2) respiration, growth, and inedible parts of the food represented mucus secreted by the animals. The assimilation of food given to the worms was figured to be 90% of the prey biomass on the basis of the analyses of Birge & Juday (1922) which showed that the crude fiber content of the sort of animal that the Phagocata were fed was about 10%. The results showed that Phagocata secreted an amount of energy in the form of mucus that was nearly equal to their body caloric content each month (0.94 ± 0.10 cal mucus/cal/month).

It is possible with the foregoing data to construct the energy balance sheet for the Phagocata (Table 7). Energy available from cannibalism was taken as one-half of the decrease in population of Phagocata. There are no other macroscopic animals which will eat flat-worms (Hyman 1951); therefore, all of the dead Phagocata were either eaten by their relatives or decomposed by microorganisms. It was not possible to measure mortality except by comparing the size of the standing crop in successive months. There is not enough information to explain the fluctuations in population size of these animals and no reason to assume that the rate of increase is constant throughout the year. Therefore, it is not possible to calculate mortality with Ricker's (1946) formula.

The fraction of assimilated energy transformed by Phagocata is very low, $\frac{18.7 \text{ KC transformed to heat}}{156.0 \text{ KC total assimilation}}$ or 12%. This is probably an adaptation made necessary by the large amounts of energy they lose in mucus which they constantly secrete. If, in calculating the fraction, the energy in the mucus is included with the energy transformed as energy "used" by the animals, the result is $\frac{107.8 \text{ KC}}{156.0 \text{ KC}}$ or 69%, which is much closer to the fraction of energy transformation by other animals.

Including the energy secreted in mucus with the energy transformed gives a valid basis for comparison with other animals of the community as the mucus secreted is a necessary part of a planarian's existence and demands a large proportion of the energy assimilated.

The efficiency of net production compared to assimilation is $\frac{113.4 \text{ KC net production}}{131.2 \text{ KC assimilated from sources outside the population}}$ or 87%.

Caddis Fly Larvae.—The caddis fly larvae in Root Spring, *Frenesia missa* (Milne), *F. difficilis* (Walker), *Limnophilis* sp., and *Lepidostoma* sp., are considered as a unit in the energy flow picture.

Frenesia difficilis and *Limnophilis* sp. were the species most commonly found. *Frenesia difficilis* was raised from the larva and identified. Limnophilis was not raised to adulthood and could not be determined to species. *Frenesia missa* was collected as the adult flying around the spring and was not associated with a larva in the spring. It may have not been present in Root Spring. The Lepidostoma larva was rare and also not associated with an adult.

The preferred habitat of the caddis larvae was not the spring but the springbrook and there were always more larvae in the latter place. The animals in the spring were individuals that had wandered into the pool due to their orientation to the current.

Very few of the pool larvae emerged as adult insects and none was collected in the tent trap set over the spring.

Caddis larvae in general are herbivorous although they will eat each other if crowded conditions prevail. The larvae of Frenesia feed almost entirely on roots and leaves of higher plants (Lloyd 1921) and in the spring they fed on those objects which fell into the water from the surrounding vegetation. Cannibalism among caddis larvae was not important in the present study as they were never present in numbers large enough to constitute crowding.

The population figures for the Trichoptera (Table 6) were checked by direct counting of the caddis larvae on the spring bottom. The errors are less than 10%.

All reproduction of the population occurs in the late fall and young larvae first appear in mid-winter. The increases in the number of individuals during the summer were due to immigration of larvae from the springbrook.

The caloric value of the Trichoptera was determined from one measurement which gave a value of 0.98 cal/mg fresh weight. This agrees with the data of Birge & Juday (1922) on the chemical composition of Trichoptera larvae.

The value for respiratory losses for the Trichoptera larvae was taken from the work of Fox & Baldes (1935), who found that at 10° C, larvae of *Limnophilus vittantus* consumed 0.73 mg O_2/gm fresh wt/hr. The animals used by Fox & Baldes averaged 4.6 mg. Because the average weight of the Trichoptera larvae in Root Spring was 14 mg, the average rate of respiration would be approximately 0.50 mg O_2/gm fresh wt/hr (based on the theory that rate of respiration is proportional to body surface, Zeuthen 1953).

For the calculation of energy flow it was assumed that from the middle of January to the middle of March there was no immigration. The larvae at that time were very small and did not move about much. (Error from this assumption could not exceed 1% of the total energy flow for these species.) Again from the middle of May to the middle of July there was no immigration because the outlet of the spring was blocked as far as the caddis larvae were concerned. The mortality rates were calculated for those intervals when there was no immigration and assumed to be the same for the months when larvae did enter the spring and direct calculation was not possible. The constant environmental conditions give a basis for this assumption with which the energy flow balance was constructed (Table 7).

The fraction of energy transformed by the caddis larvae was rather less than for the other populations in the spring, $\frac{67.5 \text{ KC}}{106.7 \text{ KC}}$ or 64%. The value is lower for the trichoptera in the springpool than it would be for those in the brook, since energy transformation which was not measured occurred in immigrant larvae before they entered the pool. The efficiency in terms of the energy passed on to other populations, net production over total energy flow, was 36%.

Pisidium virginicum Bourguignat and *Musculium partumeium* Say.—These fingernail clams live completely buried in the mud and feed on organic matter which they filter out of the water. While it was difficult to separate these two genera, it was believed that most of the population (Table 6) belonged to Pisidium.

The value for calories per unit weight of live tissue was obtained from the determinations of the chemical composition of Sphaerium (a fingernail clam) performed by Ivlev (1943). The respiration of these two species of molluscs was measured in three experiments with 23 animals and found to be 0.36 mg O_2/gm/hr.

The total mortality was calculated using the increase in the natural population from December to January to find a value for i in the formula for mortality. The difference in the mean populations for these two months was significant at $P = 0.05$.

The energy flow chart for the Pisidium and Musculium is given in Table 7. Respiration over total energy intake was 53%. The efficiency in passing organic matter on to other populations was 100% minus 53% or 47%.

Crangonyx gracilis Smith and *Physa* sp.—The amphipods and gastropods were relatively unimportant in the economy of the spring. *C. gracilis* feeds on all sorts of organic matter, both animal and plant but rarely kills its own prey. Physa is omnivorous. Unlike many of the animals in the spring, it breeds only during those months of the year when the length of daylight is more than 13½ hours (Jenner 1951). Table 6 gives the population of these two species.

The rates of respiration and of calories per unit weight for Physa were taken from the data for Pisidium. The data for Crangonyx were taken from one measurement of each variable: calories per unit weight equaled 0.81 cal/mg fresh wt; respiratory rate equaled 1.15 mg O_2/gm fresh wt/hr or 3.4 cal/cal fresh wt/month. The latter figure is reasonably close to the rate found for the other crustacean in the spring, Asellus, which respired 3.3 cal/cal/month.

Physa and Crangonyx respired at least 47.3 KC/m^2/yr and had a mortality of 9.9 KC/m^2/yr (the total amount built into their tissues during the year since at the end of the period both populations were practically absent). Assuming the efficiencies of these species to be similar to those of the other molluscs and crustaceans in the spring Physa and Crangonyx together respired roughly 90 KC/m^2/yr and passed 30 KC/m^2/yr. to other populations in the system (Table 7).

ENERGY-EXCHANGE BETWEEN THE SPRING COMMUNITY AND SURROUNDING AREAS

There were several forms of energy exchange between the spring and its surroundings: (1) dissolved and particulate organic matter contained in the water which entered and left the springpool, (2) the organic matter that entered the spring in the form of leaves, other pieces of vegetation, and animals that fell into the water, (3) the adult insects that left the system when they emerged (there was no other emigration), (4) the immigration of caddis larvae, and (5) the sunlight which was used by algae for photosynthesis. (Since the heat from sunlight was of no use to the spring organisms, it was not considered.) Of these, the energy of the emergence and immigration of insects has already been calculated.

The results of determinations of the organic content of the water entering and leaving the spring showed no significant difference in kind or amount between the organic matter being carried into the system and that being carried out. Rain did not affect these determinations as surface water did not drain into the spring and there were no noticeable short-term changes in ground water flow following rains or drought.

The most important source of energy for the spring community consisted of the leaves and other plant material that fell into the water from the surrounding land. This occurred mostly in the autumn and came to approximately 2350 kilocalories/m^2 during the year under consideration.

MICRO-ORGANISM METABOLISM

The rates of respiration given for the microflora and microfauna were obtained from the total respiration of the benthos minus the calculated rates of respiration of the known average biomass of macrofauna (Table 8). There were two periods of maximum activity of micro-organisms in Root Spring, one early in spring and the other in autumn.

TABLE 8. Photosynthesis and respiration of micro-organisms in Root Spring, Concord, Mass.

Month	Mean rate of photosynthesis	Mean rate of respiration
	$KC/m^2/month$	$KC/m^2/month$
November 1953	...	17
December	...	17
January 1954	8	33
February	56	50
March	138	25
April	250	14
May	102	25
June	68	25
July	45	20
August	23	42
September	11	50
October	10	32
	710 $KC/m^2/yr.$	350 $KC/m^2/yr.$

The total respiration of these micro-organisms was 350 $KC/m^2/yr$. A rough estimation of the portion of this amount which represents the respiration of algae may be obtained by adding three-fourths of the respiration in April when the algae were most active to one-half of the respiration in March, May, June and July; this gives 55 $KC/m^2/yr$. The respiration of micro-organisms was lowest in April and, since the algae were most active then, most of this respiration must have been due to algae. Since the algae were also active in the four other months mentioned, they must have accounted for a considerable portion of the respiration then as well.

Micro-organism metabolism throughout the year was compared with the various sources of energy upon which these organisms could have drawn. Fig. 5 gives a picture of relative rates of photosynthesis in the spring and respiration of micro-organisms and the energy available to micro-organisms from animal sources (non-predatory mortality and prey killed but not assimilated by carnivores; the chitin in cast larval and pupal skins was not included). It is apparent from the figure that the respiration of micro-organisms is much more closely correlated with the variation in usable energy lost by the macrofauna. This would indicate that the micro-organisms were probably mostly bacteria and fungi and not algae-eating animals.

The nematodes and protozoa, although not abundant, were the only other micro-organisms observed with any frequency. They probably fed mostly upon bacteria. The two genera of nematodes identified had only bacteria in their guts and members of these genera are reported to feed only upon bacteria (Nielsen 1949).

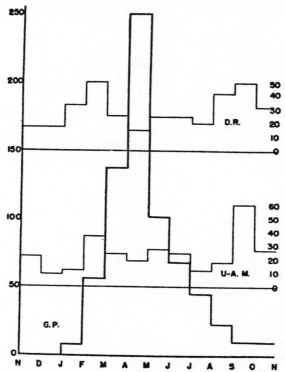

FIG. 5. Comparison of Gross Photosynthesis (G.P.); Mortality of macroscopic animals which is not assimilated by macroscopic carnivores (Unassimilated Mortality, U-A.M.); and Respiration of all micro-organisms (D.R.) for Root Spring, Concord, Mass. in 1953-54.

PHOTOSYNTHESIS

The amount of photosynthesis occurring in the spring was negligible in November and December and very slight throughout the autumn and in January, due to lack of light at those times. It was not until February that the duration and strength of the illumination became sufficient for a large crop of algae to grow in the spring. In February all of the solid surfaces in the water became covered by the filamentous green alga Stigeoclonium. This alga increased in amount and a scum of diatoms became visible to the naked eye on the surface of the sandy part of the bottom as the illumination increased. The occurrence of maximum photosynthesis of the algae in April (Table 8) is explained by the fact that after April the terrestrial plants around the spring leafed out and shaded the water. By summer this shading greatly reduced photosynthesis.

The gross production of the algae was 710 KC/m²/yr and since the algal respiration was estimated at 55 KC, net algal production was approximately 655 KC/m²/yr.

A possible source of error in these values lay in the fact that the measurements extended over 24 to 32 hours. It is possible that in this time the available phosphate or some other essential nutrient was exhausted and that the measured rate of photosynthesis was lower than the true value. However, one experiment in May was allowed to run for five days while another ran only one day in the same week and there was no significant difference in the results. Perhaps these algae are able to use organic phosphate for their growth as Rice (1949) reported for Chlorella but failed to find for Nitzschia.

BACTERIAL GROWTH ON GLASS PLATES

The numbers of bacteria which grew on the exposed parts of glass slides stuck in the middle of Root Spring allow a rough comparison to be made with other aquatic areas. One slide in the spring for 20 days showed a growth of 625 bacteria/mm²/day (21 fields counted). Two others in the water for 30 days showed 847 and 1070 bacteria/mm²/day (45 and 56 fields counted). Slides placed in a pond into which the springbrook flowed but where the water temperature was about 25° C showed more growth in the pond in two weeks than occurred in the spring in a month.

Comparison may be made with Silver Springs, Florida (Odum 1957) where slide counts at the mud surface gave 3280 bacteria mm²/day, about three times the Root Spring value. Since Silver Springs has a temperature of 23° C, it would be expected that growth there would be two to three times as fast as in Root Spring with a temperature of 9° C.

Probably some of the bacteria in Root Spring were photosynthetic but the slides were not examined for colored forms. Henrici (1939) indicates that photosynthetic bacteria are more common in shallow water than elsewhere and all of the slides in the spring were in full daylight.

Since respiration of the bacteria is included with that of the micro-organisms and their importance to the community metabolism thus evaluated, they were not considered further.

PREDATOR-PREY BALANCE

From the data for Anatopynia and Phagocata the amount of energy assimilated by carnivores each month can be found. However, a larger biomass of prey is killed than is assimilated. It has been seen that Anatopynia assimilated only about one-third of the amount that it kills and that Phagocata assimilated about 90% of its food on the average, as 10% is indigestible. Table 9 lists under "Predation" the biomass of killed prey, given in energy terms, needed to support the carnivores.

The mortalities from all causes of the various prey species are added in the first column of Table 9. Anatopynia and Phagocata are not included as prey, even though they did feed on members of their own species, because cannibalism has already been taken into account in the energy flow balances.

By comparing the total mortality of the herbivores with the predation it can be seen that for the trophic level as a whole the non-predatory mortality is 19.4%

TABLE 9. Comparison of herbivore mortality and loss due to predation (KC/m^2) in Root Spring, Concord, Mass.

Month	Mortality	Predation	Difference Non-pred. mortality
November 1953	24.0	16.8	7.2
December	21.4	20.6	0.8
January 1954	39.4	37.1	2.3
February	63.0	51.5	11.5
March	32.9	22.1	10.8
April	20.8	16.3	4.5
May	33.2	29.6	3.6
June	32.7	22.7	10.0
July	23.1	22.9	0.2
August	53.8	54.1	−0.3
September	93.8	58.7	35.1
October	29.8	24.6	5.2
	467.9	377.0	90.9

of the total mortality $\left(\dfrac{90.9}{467.9}\right)$

It seems logical to believe that non-predatory mortality would have been smaller in the spring than in other aquatic or terrestrial soils since the constant environmental conditions would result in fewer non-predatory deaths in the spring than would be the case in environments subject to freezing, anaerobiosis, etc.

It may be concluded, then, that non-predatory mortality will be one-fifth or more of the total mortality of a species or trophic level in many communities. It is not, therefore, safe to neglect this part of the energy flow through a trophic level as was done by Lindeman (1942).

RELATIVE IMPORTANCE OF VARIOUS SPECIES

An energy balance sheet for the herbivores and carnivores is given in Table 10. From this table we can compare the various groups of herbivores with respect to the amount of energy which they assimilated and the amount of energy which they transformed to heat. The oligochaetes assimilated more energy than any other group and transformed more than any other group with the exception of the isopods. The Calopsectra, which had by far the largest biomass in the summer, were third in importance in amount of energy assimilated and transformed.

These data emphasize the difficulty in making valid comparisons of the metabolic importance of various kinds of animals from subjective impressions, counts or measurements of biomass, or even energy flow determinations if these are confined to only one season. Statements of the relative importance of various kinds of animals to a community have, however, often been based on just such evidence as individual size and apparent abundance at one time of year. Indeed, the entire scheme of classifying animals as major and minor influents in a community (Clements & Shelford 1939) seems based mostly on the in-

TABLE 10. Balance sheet for herbivores and carnivores in Root Spring, Concord, Mass., 1953-54. Data for Anatopynia are divided between both trophic levels as explained in the text.

	Assimilation	Respiration
Herbivores	$KC/m^2/yr.$	$KC/m^2/yr.$
Limnodrilus	644	484
Asellus	604	486
Calopsectra	520	390
Anatopynia	159	138
Pisidium	173	91
Trichoptera larvae	88	67
Physa & Crangonyx	120	90
Total Herbivores	2300	1746
Carnivores		
Phagocata	131	19
Anatopynia	77	70
Total Carnivores	208	89

adequate criterion of individual size. The use of size as a criterion of community importance has been criticized by E. P. Odum (1953) who suggests that total biomass of a species is a better indicator of the importance of various animals in a community. An even better notion of the relative influence of different populations in an ecosystem is obtained from a study of energy flow.

The planarians were the most important carnivores even though they transformed less energy than the Anatopynia because the latter obtained more energy from plant than animal sources.

By combining all of the data so far presented, the community energy balance chart (Fig. 6) was constructed.

On the credit side of the energy balance is:

Organic debris	2350 KC/m^2/yr	(76.1%)
Gross photosynthetic production	710	(23.0%)
Immigration of caddis larvae	18	(0.6%)
Decrease in standing crop	8	(0.3%)
	3086	

This total is divided in the following way on the debit side of the balance:

Transformation to heat	2185 KC/m^2/yr	(71%)
Deposition	868	(28%)
Emigration of adult insects	33	(1%)
	3086	

PORTION OF ENERGY TRANSFORMED TO HEAT

The fraction of assimilated energy which various groups of animals in Root Spring transformed to heat during the year studied is summarized in Table 11. The fraction is close to 50% for all groups except the clams and planarians. (The value for carnivores is low because of the influence of the planarians.)

It was suggested that the reason for the very low ratio of energy transformation in the planarians is their necessity, in an evolutionary sense, to compensate for their large loss of energy in the form of mucus. However, since the energy in mucus serves the worms in their activities and is subsequently lost to them just as the energy used for doing work and transformed to heat is used and lost, the energy in mucus may logically be included with the energy transformed to heat in calculating the ratio. The portion of energy transformed plus mucus energy was 69% which is comparable to the value for the other animals. The relatively low transforming ratio of the clams may possibly have a similar explanation since these animals lose considerable mucus in their pseudofeces.

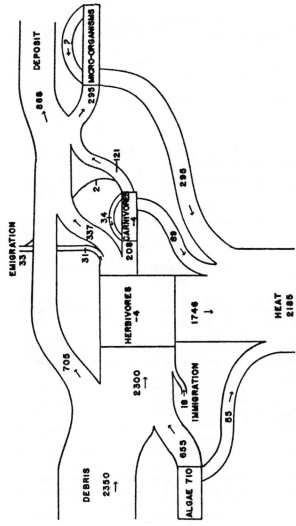

Fig. 6. Energy flow diagram for Root Spring, Concord, Mass. in 1953-54. Figures in KC/m²/yr; numbers inside boxes indicate changes in standing crops; arrows indicate direction of flow.

TABLE 11. Energy ratios for various groups of organisms from Root Spring and from other sources. Efficiencies for Mytilus were calculated for season of maximum growth and would be lower if figures on an annual basis as were Root Spring data.

	Net Production / Assimilation	Transformed to Heat / Assimilation
Root Spring		
Calopsectra	20%	80%
Anatopynia	13	84
Limnodrilus	26	74
Asellus	20	80
Trichoptera larvae	36	64
Pisidium	47	53
Planarians	87	12
All herbivores	25	75
All carnivores	59	37
Entire Community	29	71
Jorgensen (1952)		
Veligers, marine	60 - 70	..
Mytilus, 40.9-49.0mm	54	..
Mytilus, 90mm	11	..
Harvey (1950)		
Calanus	70	..
Brody (1945)		
Embryos	50 - 65	..

EFFICIENCY OF NET PRODUCTION

The efficiency of net production (Table 11) varied around 30% for the animals in the spring except for the clams and planarians with efficiencies of 47% and 87% respectively. These latter two values are close to those for larval animals and embryos while values from 20-30% are more common for efficiencies of post natal growth (Jørgensen 1952). It is possible that the high efficiency in the planarians is associated with their exceptional regenerative ability.

The net production efficiency for the entire community was 29%. In other words, there would be a continual accumulation of usable energy in the ecosystem if organic matter were not removed.

A small amount of organic matter was removed in the bodies of the insects that emerged, 33 KC/m^2, in comparison to the deposition of 868 KC/m^2. The pool was prevented from filling completely by shifts in the position of the boils which stir up and wash out some mud.

COMPARISON WITH DIFFERENT COMMUNITIES

With the data available in the literature comparisons can be made, not only of the primary productivity in a series of communities but also of the energy flow through the trophic levels of a much smaller number of systems.

Table 12 gives values for energy available to organims other than primary producers (primary net production plus accumulated energy in organic matter) and the ratio of this to total incident light. Incident light in the Concord region was calculated to be 1.095×10^6 KC/m²/yr from weather bureau data (U.S. Dept. of Commerce 1951). In aquatic communities some light is absorbed by the water but in order to make comparisons between land and water systems as producers on the earth, no correction was made.

TABLE 12. Efficiencies and productivities of various communities compared with those of Root Spring, Concord, Mass.

	Energy available to consumers	Column One Light	Gross Prod. Light	Source
	KC/m²/yr.			
Root Spring	3005	0.27%	0.2%*	
Eniwetok Reef	21800	1.8	5.8	Odum & Odum 1955
Cedar Bog Lake	879	0.074	0.1	Lindeman 1942
Lake Mendota**	3730	0.31	0.4	Lindeman 1942
Minnesota Pond	394	0.033	0.04	Dineen 1953
Corn field in summer	6170	1.2	1.6	Transeau 1926
Georges Bank	.....	.....	0.3	Clarke 1946
Silver Spring, Florida	.....	.....	8.0	Odum 1957
Average Terrestrial plant	.....	.....	0.09	Riley 1944
"Best forests"	.....	.....	0.25	Riley 1944

*For April only.
**Values considered erroneously high by Lindeman.

Root Spring provided more energy per unit area for consumers than average temperate communities because of its action in accumulating energy that had been fixed by plants outside of the spring. The spring community itself expended no energy to bring about this accumulation.

As far as primary gross production is concerned, the algae of Root Spring were as efficient as the better temperate communities. The efficiency was calculated for April before shading by trees was important, since no direct measurement of light intensity was made.

Comparisons of the different trophic levels of a few communities are made in Table 13. The ratio, $\frac{\text{assimilation of one trophic level}}{\text{net production of next lower level}}$, is used for comparison. This represents the efficiency of utilization of available food by a trophic level. Values were calculated from the data given by the various authors.

The most striking difference shown in Table 13 between the various communities is the large assimilation and efficiency of herbivores in Root Spring. (The large carnivore assimilation appears to be a result of the large food supply in the form of herbivores since the efficiency of carnivores in Root Spring was about the same as in the other communities.)

The reason that a large portion of the available plant energy was assimilated probably lies in the fact that the Root Spring conditions were more favorable to life than those in the other environments of Table 13. For example, water currents in the spring provided sufficient oxygen and rapid removal of metabolic wastes at all times. It must be noted, however, that the transformation by a community of as much or more than three-fourths of the incoming usable energy to heat is not unusual.

If in any community the primary net production is not accumulating to any appreciable extent, as is the case in the woods in the south of England (Pearsall 1948), it is because the efficiency of utilization of primary net production is close to 100%. In other words, if the primary consumers consume all the primary net production there will be no accumulation of organic matter in the community. This contradicts Lindeman's (1942) thesis that the utilization efficiency will be progressively greater for higher trophic levels.

In Root Spring the primary consumers are mostly macroscopic animals. These animals are herbivores, detritus feeders, and scavengers. In terrestrial soils, on the other hand, the most important primary consumers are probably microscopic, heterotrophic plants. Data summarized from Stöckli (1946) indicate that the microflora—fungi, bacteria, actinomycetes (and algae)—compose most of the biomass of terrestrial soil organisms, and since these organisms are also very small, they are undoubtedly the most active metabolically. These plants are "decomposers," but are still primary consumers as long as they are burning fuel from green plant sources. It makes no difference in an energy flow analysis, whether organisms which obtain energy from plants are herbivores, saprophytes, detritus feeders or decomposers; they are still all primary consumers.

A complicating factor arises from the fact that decomposers, etc. feed upon consumers as well as producers (i.e. are secondary and tertiary as well as primary consumers). This complication is common to all soil communities, aquatic and terrestrial, and does not invalidate the comparisons made from Table 13.

The data from this investigation raise a number

TABLE 13. Efficiencies and assimilations for higher trophic levels of various communities compared with those of Root Spring, Concord, Mass.

	ROOT SPRING		CEDAR BOG LAKE		LAKE MENDOTA		MINNESOTA POND		SILVER SPRING
	Assim.*	Effic.**	Lindeman (1942) Assim.	Effic.	Lindeman (1942) Assim.	Effic.	Dineen (1953) Assim.	Effic.	Odum 1957 Assim.
Primary Consumers (Herbivores)	2300	76%	148	16.8%	416	11.2%	92	23%	1280
Secondary Consumers (Carnivores)	208	36%	31	29.8%	23	8.7%	34	47%	28
Tertiary Consumers (Secondary Carnivores)	....	....	....	......	3	23.0%	..	....	....

*All assimilations are KC/m²yr.   Assimilation
**All efficiencies are ───────── Net production of next lower trophic level

of interesting questions: Is it a general rule that the ratio of energy flow of secondary consumers to that of primary consumers remains about the same for all communities? Does the stability of a community affect the aforementioned ratio? What relation might there be between assimilation by primary consumers and producer net production? Is there a correlation between body size of consumers and deposit or removal of a sizable fraction of primary net production? These and other questions can only be answered when more data from both experimental and natural communities are available.

SUMMARY

1. A study of community metabolism in terms of energy flow was undertaken in 1953-54 in order to provide a more accurate picture of energy flow through the populations of a community than had hitherto been available. Root Spring, a temperate cold spring in Concord, Massachusetts, was chosen for study because of the relatively constant environmental conditions and simple biota.

2. The energy flow through the larger animal populations was studied in detail. These animals included two chironomids, *Calopsectra dives* and *Anatopynia dyari*, a tubificid oligochaete, an isopod, two planarian species, three species of caddis larvae, two species of fingernail clams, one snail, and one amphipod. To minimize artificial disturbance of the community the animals in the samples were kept alive and returned to the spring after being weighed and counted.

3. Respiration of the larger species of animals was measured in containers filled with spring water and placed in the spring so that conditions were as nearly natural as possible. Photosynthesis and respiration of the micro-organisms were determined by measuring oxygen changes in water over bottom mud enclosed in light and dark, covered glass cylinders.

4. Non-predatory mortality was calculated as the difference between the total mortality determined with the formulae published by Ricker (1946) and the mortality due to predation. Non-predatory mortality accounted for nearly 20% of the mortality occurring in the herbivore trophic level. Reasons were given for believing that this was a lower rate of non-predatory mortality than exists in most communities and that non-predatory mortality may not be safely neglected in studies of productivity.

5. The Anatopynia were found to assimilate only 30% of the prey which they killed. This carnivorous midge larva actually obtained about two-thirds of its food from plant sources which enabled it to exist during periods in which it could not obtain sufficient animal food.

6. Phagocata, the planarian, lost more energy in the mucus which it secreted than it used in doing external and internal work as measured by respiration. **Mucus secretion took 79%** of this species' net production.

7. 76% of the energy transformed by the organisms entered the spring in leaves, fruit and branches of terrestrial vegetation. Photosynthesis in the spring accounted for 23% and less than 1% entered in the bodies of immigrating caddis larvae.

8. The micro-organisms respired about 350 KC/m^2/yr. Of this only 55 KC were respired by algae, the remainder by bacteria, fungi, protozoa and nematodes. Evidence was given to indicate that most of the micro-organisms were feeding upon animal remains.

9. The most conspicuous group of herbivores, Calopsectra, was only third in importance in energy flow. It was suggested that the relative amount of energy which flows through each population is a better criterion than size or biomass of the importance of various animal species in a community.

10. The energy balance sheet for the community shows that, of the year's total energy input, 71% was transformed to heat, 28% was deposited in the community and only 1% emerged in adult insects.

11. The efficiency of energy transformation, energy used for internal and external work as measured by respiration divided by assimilation, was close to 80% for all animals but the planarians and clams. The clams transformed only 53% of their assimilation, the planarians only 12%. These low rates of respiration may be an adaptation connected with the large amounts of mucus lost by these animals.

12. The ratio $\dfrac{\text{assimilation}}{\text{net production of algae plus energy of organic debris}}$ was 76% in Root Spring, higher than other aquatic communities investigated but probably not as high as for some soils.

13. The efficiency $\dfrac{\text{algal photosynthesis}}{\text{available solar energy}}$ in Root Spring in April when shading of the water at a minimum was 0.2% which is comparable to any temperate area. If the efficiency of the spring community is calculated on the basis of total inflow of energy instead of just photosynthesis, the efficiency for the year was 0.27%, higher than for most temperate latitude communities which have been studied.

14. The assimilation by macroscopic primary consumers (over 1 mg) was five times that reported for any other temperate latitude aquatic community. The steady flow of constant temperature water bring-

ing oxygen, removing waste, and enabling the animals to be active all year was probably responsible for this high assimilation.

LITERATURE CITED

Birch L. C. & D. P. Clark. 1953. Forest soil as an ecological community with special reference to the faunas. Quart. Rev. Biol. **28**: 13-36.

Birge, E. A. & C. Juday. 1922. The inland lakes of Wisconsin. The plankton, Part I: Its quantity and chemical composition. Bull. Wis. Geol. Nat. Hist. Surv. **64**: 1-222.

Borutzky, E. V. 1939. (Dynamics of *Chironomus plumosus* in the profundal of Lake Beloie.) Proc. Kossino Limnol. Stat. **22**: 156-195. (Russian with English summary p. 190-195.)

Brody, S. 1945. Bioenergetics and growth. New York: Reinhold Publ. Corp. 1023 pp.

Brues, C. T. 1928. Studies on the fauna of hot springs in the Western United States and the biology of thermophilous animals. Proc. Am. Acad. Arts Sci. **63**: 139-228.

Bullock, T. H. 1955. Compensation for temperature in the metabolism and activity of poikilotherms. Biol. Rev. **30**: 311-342.

Clarke, G. L. 1946. Dynamics of production in a marine area. Ecol. Monog. **16**: 321-335.

Clements, F. E., & V. E. Shelford 1939. Bio-ecology. New York: John Wiley & Sons. 425 pp.

Dineen, C. F. 1953. An ecological study of a Minnesota pond. Am. Midland Nat. **50**: 349-376.

Dudley, Patricia L. 1953. A faunal study of four springbrooks in Boulder County, Colorado. Master's thesis, Univ. of Colorado.

Ewer, R. F. 1941. On the function of haemoglobin in *Chironomus*. J. Exper. Biol. **18**: 13-205.

Fox, H. M., & E. J. Baldes. 1935. Quoted in Prosser, C. L., 1950. Comparative Animal Physiology. Philadelphia: W. B. Saunders Co.

Fox, H. M., & C. A. Wingfield. 1938. A portable apparatus for the determination of oxygen dissolved in a small volume of water. J. Exper. Biol. **15**: 437-445.

Gerking, S. D. 1954. The food turnover of a bluegill population. Ecology **35**: 490-497.

Harvey, H. W. 1950. On the production of living matter in the sea off Plymouth. J. Mar. Biol. Ass. U.K. **29**: 97-137.

Henrici, A. T. 1936. Studies of freshwater bacteria. III. Quantitative aspects of the direct microscopic method. J. Bact. **32**: 265-280.

Hyman, L. H. 1919. Oxygen consumption in relation to feeding and starvation. Amer. Jour. Physiol. **49**:

377-402.

Hyman, L. H. 1951. The Invertebrates. Vol. II. Platyhelminthes and Rhynchocoela, the acoelomate Bilateria. New York: McGraw-Hill Book Co. 550 pp.

Ivlev, V. S. 1934. Eine Mikromethode zur Bestimmung des Kaloriengehalts von Nahrstoffen. Biochem. Ztschr. **275**: 49-55.

Ivlev, V. S. 1945. The biological productivity of waters (Translation by W. E. Ricker) Adv. Mod. Biol. **19**: 98-120.

Jenner, C. E. 1951. Photoperiodism in the fresh-water pulmonate snail, *Lymnaea palustris*. Ph.D. thesis, Harvard University.

Johannsen, O. A. 1937. Aquatic Diptera Pt. III Chironomidae: Subfamilies Tanypodinae, Diamesinae, and Orthocladiinae. Mem. Cornell Univ. Agr. Exp. Sta. **205**: 1-84.

Jørgensen, C. B. 1952. Efficiency of growth in *Mytilus edulis* and two gastropod veligers. Nature **170**: 714-715.

Lindeman R. L. 1942. The trophic-dynamic aspect of ecology. Ecology **23**: 399-418.

Lloyd J. T. 1921. The biology of North American caddis fly larvae. Bull. Lloyd Libr. **21**: Ento. Ser. No. 1.

Macfadyen, A. 1948. The meaning of productivity in biological systems. J. Anim. Ecol. **17**: 75-80.

Nielsen, C. Overgaard, 1949. Studies on the soil microfauna. II. The soil inhabiting nematodes. Natura Jutlandica **2**: 1-131.

Odum, E. P. 1953. Fundamentals of Ecology. Philadelphia: W. B. Saunders Co. 384 pp.

Odum, H. T. 1956. Efficiencies, size of organisms, and community structure. Ecology **37**: 592-597.

Odum, H. T. 1957. Trophic structure and productivity of Silver Springs, Florida. Ecol. Monog. **27**: 55-112.

Odum, H. T. & E. P. Odum. 1955. Trophic structure and productivity of a windward coral reef community on Eniwetok Atoll. Ecol. Monog. **25**: 291-320.

Pearsall, W. H. 1948. An ecologist looks at soil. Symposium on Soil Biology, British Ecological Society, April, 1948.   J. Ecol. **36**: 328.

Pennak, R. W. 1946. Annual limnological cycles in some Colorado reservoir lakes. Ecol. Monog. **19**: 233-267.

Pennak, R. W. 1954. Fresh-water invertebrates of the United States. New York: Ronald Press. 769 pp.

Rice, T. 1949. The effects of nutrients and metabolites on populations of planktonic algae. PhD. thesis, Harvard University.

Ricker, W. E. 1946. Production and utilization of fish populations. Ecol. Monog. **16**: 375-389.

Riley, G. A. 1944. The carbon metabolism and photosynthetic efficiency of the earth as a whole. Amer. Sci. **32**: 129-134.

Stöckli, A. 1946. Die biologische Komponente der Vererdung, der Gare und der Nährstoffpufferung. Schweiz. Landw. Monatsh. **24**.

Transeau, E. N. 1926. The accumulation of energy by plants. Ohio J. Sci. **26**: 1-10.

U. S. Department of Commerce. 1951. Weather Bureau, climatological data for the U. S. by sections. Vol. **38**. Washington.

Walshe-Maetz, B. M. 1953. Le metabolisme de *Chironomus plumosus* dans des conditions naturelles. Physiol. Comp. e. Ecologia **111**: 135-154.

Welch, P. S. 1948. Limnological Methods. Philadelphia: Blakiston Press. 381 pp.

Zeuthen, E. 1953. Oxygen uptake as related to body size in organisms. Quart. Rev. Biol. **28**: 1-12.

Concluding Remarks

G. Evelyn Hutchinson

This concluding survey[1] of the problems considered in the Symposium naturally falls into three sections. In the first brief section certain of the areas in which there is considerable difference in outlook are discussed with a view to ascertaining the nature of the differences in the points of view of workers in different parts of the field; no aspect of the Symposium has been more important than the reduction of areas of dispute. In the second section a rather detailed analysis of one particular problem is given, partly because the question, namely, the nature of the ecological niche and the validity of the principle of niche specificity has raised and continues to raise difficulties, and partly because discussion of this problem gives an opportunity to refer to new work of potential importance not otherwise considered in the Symposium. The third section deals with possible directions for future research.

The Demographic Symposium as a Heterogeneous Unstable Population

In the majority of cases the time taken to establish the general form of the curve of growth of a population from initial small numbers to a period of stability or of decline is equivalent to a number of generations. If, as in the case of man, the demographer is himself a member of one such generation, his attitude regarding the nature of the growth is certain to be different from that of an investigator studying, for instance, bacteria, where the whole process may unfold in a few days, or insects, where a few months are required for several cycles of growth and decline. This difference is apparent when Hajnal's remarks about the uselessness of the logistic are compared with the almost universal practice of animal demographers to start thinking by making some suitable, if almost unconscious, modification of this much abused function.

[1] I wish to thank all the participants for their kindness in sending in advance manuscripts or information relative to their contributions. All this material has been of great value in preparing the following remarks, though not all authors are mentioned individually. Where a contributor's name is given without a date, the reference is to the contribution printed earlier in this volume. I am also very much indebted to the members (Dr. Jane Brower, Dr. Lincoln Brower, Dr. J. C. Foothills, Mr. Joseph Frankel, Dr. Alan Kohn, Dr. Peter Klopfer, Dr. Robert MacArthur, Dr. Gordon A. Riley, Mr. Peter Wangersky, and Miss Sally Wheatland) of the Seminar in Advanced Ecology, held in this department during the past year. Anything that is new in the present paper emerged from this seminar and is not to be regarded specifically as an original contribution of the writer.

The human demographer by virtue of his position as a slow breeding participant observer, and also because he is usually called on to predict for practical purposes what will happen in the immediate future, is inevitably interested in what may be called the microdemography of man. The significant quantities are mainly second and third derivatives, rates of change of natality and mortality and the rates of change of such rates. These latter to the animal demographer might appear as random fluctuations which he can hardly hope to analyse in his experiments. What the animal demographer is mainly concerned with is the macrodemographic problem of the integral curve and its first derivative. He is accustomed to dealing with innumerable cases where the latter is negative, a situation that is so rare in human populations that it seems to be definitely pathological to the human demographer. Only when anthropology and archaeology enter the field of human demography does something comparable to animal demography, with its broad, if sometimes insufficiently supported generalisations and its fascinating problems of purely intellectual interest, emerge. From this point of view the papers of Birdsell and Braidwood are likely to appeal most strongly to the zoologist, who may want to compare the rate of spread of man with that considered by Kurtén (1957) for the hyena.

It is quite likely that the difference that has just been pointed out is by no means trivial. The environmental variables that affect fast growing and slow growing populations are likely to be much the same, but their effect is qualitatively different. Famine and pestilence may reduce human populations greatly but they rarely decimate them in the strict sense of the word. Variations, due to climatic factors, of insect populations are no doubt often proportionately vastly greater. A long life and a long generation period confer a certain homeostatic property on the organisms that possess them, though they prove disadvantageous when a new and powerful predator appears. The elephant and the rhinoceros no longer provide models of human populations, but in the early Pleistocene both may have done so. The rapid evolution of all three groups in the face of a long generation time is at least suggestive.

It is evident that a difference in interest may underlie some of the arguments which have enlivened, or at times disgraced, discussions of this subject. Some of the most significant modern

COLD SPRINGS HARBOR SYMPOSIUM ON QUANTITATIVE BIOLOGY, 1957, Vol. 22, pp. 415-427.

work has arisen from an interest in extending the concepts of the struggle for existence put forward as an evolutionary mechanism by Darwin practically a century ago. Such work, of which Lack's recent contributions provide a distinguished example, tends to concentrate on relatively stable interacting populations in as undisturbed commmunition as possible. Another fertile field of research has been provided by the sudden increases in numbers of destructive animals, often after introduction or disturbance of natural environments. Here more than one point of view has been apparent. Where emphasis has been on biological control, that is, a conscious rebuilding of a complex biological association, a view point not unlike that of the evolutionist has emerged—where emphasis has been placed on the actual events leading to a very striking increase or decrease in abundance, given the immediate ecological conditions, the latter have appeared to be the most significant variables. Laboratory workers have moreover tended to keep all but a few factors constant, and to vary these few systematically. Field workers have tended to emphasize the ever changing nature of the environment. It is abundantly clear that all these points of view are necessary to obtain a complete picture. It is also very likely that the differences in initial point of view are often responsible for the differences in the interpretation of the data.

The initial differences of point of view are not the only difficulty. In the following section an analysis of a rather formal kind of one of the concepts frequently used in animal ecology, namely that of the *niche*, is attempted. This analysis will appear to some as compounded of equal parts of the obvious and the obscure. Some people however may find when they have worked through it, provided that it is correct, that some removal of irrelevant difficulties has been achieved. It is not necessary in any empirical science to keep an elaborate logicomathematical system always apparent, any more than it is necessary to keep a vacuum cleaner conspicuously in the middle of a room at all times. When a lot of irrelevant litter has accumulated the machine must be brought out, used, and then put away. It might be useful for those who argue that the word environment should refer to the environment of a population, and those who consider it should been the environment of an organism, to use the word both ways for a couple of months, writing "environment" when a single individual is involved, "Environment" when reference is to a population. In what follows the term will as far as possible not be used, except in the non-committal adjectival form environmental, meaning any property outside the organisms under consideration.

The Formalisation of the Niche and the Volterra-Gause Principle

Niche space and biotop space

Consider two independent environmental variables x_1 and x_2 which can be measured along ordinary rectangular coordinates. Let the limiting values permitting a species S_1 to survive and reproduce be respectively x'_1, x''_1 for x_1 and x'_2, x''_2 for x_2. An area is thus defined, each point of which corresponds to a possible environmental state permitting the species to exist indefinitely. If the variables are independent in their action on the species we may regard this area as the rectangle ($x_1 = x'_1, x_1 = x''_1, x_2 = x'_2, x_2 = x''_2$), but failing such independence the area will exist whatever the shape of its sides.

We may now introduce another variable x_3 and obtain a volume, and then further variables $x_4 \ldots x_n$ until all of the ecological factors relative to S_1 have been considered. In this way an n-dimensional hypervolume is defined, every point in which corresponds to a state of the environment which would permit the species S_1 to exist indefinitely. For any species S_1, this hypervolume $\mathbf{N}_1$ will be called the *fundamental niche*[2] of S_1. Similarly for a second species S_2 the fundamental niche will be a similarly defined hypervolume $\mathbf{N}_2$.

It will be apparent that if this procedure could be carried out, all X_n variables, both physical and biological, being considered, the fundamental niche of any species will completely define its ecological properties. The fundamental niche defined in this way is merely an abstract formalisation of what is usually meant by an ecological niche.

As so defined the fundamental niche may be regarded as a set of points in an abstract n-dimensional $\mathbf{N}$ space. If the ordinary physical space $\mathbf{B}$ of a given biotop be considered, it will be apparent that any point $p(\mathbf{N})$ in $\mathbf{N}$ can correspond to a number of points $p_i(\mathbf{B})$ in $\mathbf{B}$, at each one of which the conditions specified by $p(\mathbf{N})$ are realised in $\mathbf{B}$. Since the values of the environmental variables $x_1\, x_2 \ldots x_n$ are likely to vary continuously, any subset of points in a small elementary volume $\Delta \mathbf{N}$ is likely to correspond to a number of small elementary volumes scattered about in $\mathbf{B}$. Any volume $\mathbf{B}'$ of the order of the dimensions of the mean free paths of any animals under consideration is likely to contain points corresponding to points in various fundamental niches in $\mathbf{N}$.

Since $\mathbf{B}$ is a limited volume of physical space comprising the biotope of a definite collection of species $S_1, S_2 \cdots S_n$, there is no reason why a given point in $\mathbf{N}$ should correspond to any points in $\mathbf{B}$. If, for any species S_1, there are no points in

---

[2] This term is due to MacArthur. The general concept here developed was first put forward very briefly in a footnote (Hutchinson, 1944).

B corresponding to any of the points in N_1, then B will be said to be *incomplete* relative to S_1. If some of the points in N_1 are represented in B then the latter is *partially incomplete* relative to S_1, if all the points in N_1 are represented in B the latter is *complete* relative to S_1.

Limitations of the set-theoretic mode of expression. The following restrictions are imposed by this mode of description of the niche.

1. It is supposed that all points in each fundamental niche imply equal probability of persistance of the species, all points outside each niche, zero probability of survival of the relevant species. Ordinarily there will however be an optimal part of the niche with markedly suboptimal conditions near the boundaries.

2. It is assumed that all environmental variables can be linearly ordered. In the present state of knowledge this is obviously not possible. The difficulty presented by linear ordering is analogous to the difficulty presented by the ordering of degrees of belief in non-frequency theories of probability.

3. The model refers to a single instant of time. A nocturnal and a diurnal species will appear in quite separate niches, even if they feed on the same food, have the same temperature ranges etc. Similarly, motile species moving from one part of the biotop to another in performance of different functions may appear to compete, for example, for food, while their overall fundamental niches are separated by strikingly different reproductive requirements. In such cases the niche of a species may perhaps consist of two or more discrete hypervolumes in N. MacArthur proposed to consider a more restricted niche describing only variables in relation to which competition actually occurs. This however does not abolish the difficulty. A formal method of avoiding the difficulty might be derived, involving projection onto a hyperspace of less than n-dimensions. For the purposes for which the model is devised, namely a clarification of niche-specificity, this objection is less serious than might at first be supposed.

4. Only a few species are to be considered at once, so that abstraction of these makes little difference to the whole community. Interaction of any of the considered species is regarded as competitive in sense 2 of Birch (1957), negative competition being permissible, though not considered here. All species other than those under consideration are regarded as part of the coordinate system.

Terminology of subsets. If N_1 and N_2 be two fundamental niches they may either have no points in common in which case they are said to be *separate*, or they have points in common and are said to *intersect*.

In the latter case:

$(N_1 - N_2)$ is the subset of N_1 of points not in N_2
$(N_2 - N_1)$ is the subset of N_2 of points not in N_1

$N_1 \cdot N_2$ is the subset of points common to N_1 and N_2, and is also referred to as the *intersection subset*.

Definition of niche specificity. Volterra (1926, see also Lotka 1932) demonstrated by elementary analytic methods that under constant conditions two species utilizing, and limited by, a common resource cannot coexist in a limited system.[3] Winsor (1934) by a simple but elegant formulation showed that such a conclusion is independent of any kind of finite variations in the limiting resource. Gause (1934, 1935) confirmed this general conclusion experimentally in the sense that if the two species are forced to compete in an undiversified environment one inevitably becomes extinct. If there is a diversification in the system so that some parts favor one species, other parts the other, the two species can coexist. These findings have been extended and generalised to the conclusion that two species, when they co-occur, must in some sense be occupying different niches. The present writer believes that properly stated as an empirical generalisation, which is true except in cases where there are good reasons not to expect it to be true,[4] the principle is of fundamental importance and may be properly called the Volterra-Gause Principle. Some of the confusion surrounding the principle has arisen from the concept of two species not being able to co-occur when they occupy identical niches. According to the formulation given above, identity of fundamental niche would imply $N_1 = N_2$, that is, every point of N_1 is a member of N_2 and every point of N_2 a member of N_1. If the two species S_1 and S_2 are indeed valid species distinguishable by a systematist and not freely interbreeding, this is so unlikely that the case is of no empirical interest. In terms of the set-theoretic presentation, what the Volterra-Gause principle meaningfully states is that for any small element of the intersection subset $N_1 \cdot N_2$, there do not exist in B corresponding small parts, some inhabited by S_1, others by S_2.

Omitting the quasi-tautological case of $N_1 = N_2$, the following cases can be distinguished.

(1) N_2 is a proper subset of N_1 (N_2 is "inside" N_1)

   (a) competition proceeds in favor of S_1 in all the elements of B corresponding to $N_1 \cdot N_2$; given adequate time only S_1 survives.

   (b) competition proceeds in favor of S_2 in all elements of B corresponding to some part of the intersection subset and both species survive.

(2) $N_1 \cdot N_2$ is a proper subset of both N_1 and N_2; S_1 survives in the parts of B space

---

[3] I regret that I am unable to appreciate Brian's contention (1956) that the Volterra model refers only to interference, and the Winsor model to exploitation.

[4] cf. Schrödinger's famous restatement of Newton's First Law of Motion, that a body perseveres at rest or in uniform motion in a right line, except when it doesn't.

corresponding to ($N_1 = N_2$), S_2 in the parts corresponding to ($N_2 = N_1$), the events in $N_1 = N_2$ being as under I, with the proviso that no point in $N_1 \cdot N_2$ can correspond to the survival of both species.

In this case the two difference subsets ($N_1 - N_2$) and ($N_2 - N_1$) are, in Gause's terminology, refuges for S_1 and S_2 respectively.

If we define the realised niche N'_1 of S_1 in the presence of S_2 as ($N_1 - N_2$), if it exists, plus that part of $N_1 \cdot N_2$ as implies survival of S_1, and similarly the realised niche N'_2 of S_2 as ($N_2 - N_1$), if it exists, plus that part of $N_1 \cdot N_2$ corresponding to survival of S_2, then the Volterra-Gause principle is a statement of an empirical generalisation, which may be verified or falsified, that realised niches do not intersect. If the generalisation proved to be universally false, the falsification would presumably imply that in nature resources are never limiting.

Validity of the Gause-Volterra Principle. The set-theoretic approach outlined above permits certain refinements which, however obvious they may seem, apparently require to be stated formally in an unambiguous way to prevent further confusion. This approach however tells us nothing about the validity of the principle, but merely where we should look for its verification or falsification.

Two major ways of approaching the problem have been used, one experimental, the other observational. In the experimental approach, the method (*e.g.* Gause, 1934, 1935; Crombie, 1945, 1946, 1947) has been essentially to use animal populations as elements in analogue computers to solve competition equations. As analogue computers, competing populations leave much to be desired when compared with the more conventional electronic machines used for instance by Wangersky and Cunningham. At best the results of laboratory population experiments are qualitatively in line with theory when all the environmental variables are well controlled. In general such experiments indicate that where animals are forced by the partial incompleteness of the **B** space to live in competition under conditions corresponding to a small part of the intersection subset, only one species survives. They also demonstrate that the identity of the survivor is dependent on the environmental conditions, or in other words on which part of the intersection subset is considered, and that when deliberate niche diversification is brought about so that at least one non-intersection subset is represented in **B**, two species may co-occur indefinitely. It would of course be most disturbing if confirmatory models could not be made from actual populations when considerable trouble is taken to conform to the postulates of the deductive theory.

The second way in which confirmation has been sought, namely by field studies of communities consisting of a number of allied species also lead to a confirmation of the theory, but one which may need some degree of qualification. Most work has dealt with pairs of species, but the detailed studies on *Drosophila* of Cooper and Dobzhansky (1956) and of Da Cunha, El-Tabey Shekata and de Olivera (1957), to name only two groups of investigators, the investigation of about 18 species of *Conus* on Hawaiian reef and littoral benches (Kohn, in press) and the detailed studies of the food of six co-occurring species of *Parus* (Betts, 1955) indicate remarkable cases among many co-occurring species of insects, mollusks and birds respectively. However much data is accumulated there will almost always be unresolved questions relating to particular species, though the presumption from this sort of work is that, in any large group of sympatric species belonging to a single genus or subfamily, careful work will always reveal ecological differences. The sceptic may reply in two ways, firstly pointing out that the quasi-tautological case of $N_1 = N_2$ has already been dismissed as too improbable to be of interest, and that when a great deal of work has to be done to establish the difference, we are getting as near to niche identity as is likely in a probabilistic world. Occasionally it may be possible to use indirect arguments to show that the differences are at least evolutionarily significant. Lack (1947b) for instance points out that in the Galapagos Islands, among the heavy billed species of *Geospiza*, where both *G. fortis* Gould, and *G. fuliginosa* Gould co-occur on an island, there is a significant separation in bill size, but where either species exists alone, as on Crossman Island and Daphne Island the bills are intermediate and presumably adapted to eating modal sized food. This is hard to explain unless the small average difference in food size believed to exist between sympatric *G. fortis* and *G. fuliginosa* is actually of profound ecological significance. The case is particularly interesting as most earlier authors have dismissed the significance of the small alleged differences in the size of food taken by the species. Few cases of specific ecological difference encountered outside *Geospiza* would appear at first sight so tenuous as this.

A more important objection to the Volterra-Gause principle may be derived from the extreme difficulty of identifying competition as a process actually occurring in nature. Large numbers of cases can of course be given in which there is very strong indirect evidence of competitive relationships between species actually determining their distribution. A few examples may be mentioned. In the British Isles (Hynes, 1954, 1955) the two most widespread species of *Gammarus* in freshwater are *Gammarus deubeni* Lillj; and *G. pulex* (L.). The latter is the common species in England and most of the mainland of Scotland, the former is found exclusively in Ireland, the Shetlands, Orkneys and most of the other Scottish Islands and in Cornwall. On northern mainland Scotland only

G. lacustris Sars. is found. Both *deubeni* and *pulex* occur on the Isle of Man and in western Cornwall. Only in the Isle of Man have the two species been taken together. It is extremely probable that *pulex* is a recent introduction to that island. *G. deubeni* is well known in brackish water around the whole of northern Europe. It is reasonable to suppose that the fundamental niches of the two species overlap, but that within the overlap *pulex* is successful, while *deubeni* with a greater tolerance of salinity has a refuge in brackish water. Hynes moreover shows that *G. pulex* has a biotic (reproductive) potential two or three times that of *deubeni* so that in a limited system inhabitable by both species, under constant conditions *deubeni* is bound to be replaced by *pulex*. This case is as clear as one could want except that Hynes is unable to explain the absence of *G. deubeni* from various uninhabited favorable localities in the Isle of Man and elsewhere. Hynes also notes that Steusloff (1943) had similar experiences with the absence of *Gammarus pulex* in various apparently favorable German localities. Ueno (1934) moreover pointed out that *Gammarus pulex* (*sens. lat.*) occurs abundantly in Kashmir up to 1600 meters, and is an important element in the aquatic fauna of the Tibetan highlands to the east above 3800 miles, but is quite absent in the most favorable localities at intermediate altitudes. These disconcerting empty spaces in the distribution of *Gammarus* may raise doubts as to the completeness of the picture presented in Hynes' excellent investigations.

Another very well analysed case (Dumas, 1956) has been recently given for two sympatric species of *Plethodon*, *P. dunni* Bishop, and *P. vehiculum* (Cooper), in the Coastal Ranges of Oregon. Here experiments and field observations both indicate that *P. dunni* is slightly less tolerant of low humidity and high temperature than is *P. vehiculum*, but when both co-occur *dunni* can exclude *vehiculum* from the best sites. However under ordinary conditions in nature the number of unoccupied sites which appear entirely suitable is considerable, so that competition can not be limiting except in abnormally dry years.

In both these cases, which are two of the best analysed in the literature, the extreme proponent of the Volterra-Gause principle could argue that if the investigator was equipped with the sensory apparatus of *Gammarus* or *Plethodon* he would know that the supposedly suitable unoccupied sites were really quite unsuitable for any self respecting member of the genus in question. This however is pure supposition.

Even in the rather conspicuous case of the introduction of *Sciurus carolinensis* Gmelin and its spread in Britain, the popular view that the bad bold invader has displaced the charming native *S. vulgaris leucourus* Kerr, is apparently mythological. Both species are persecuted by man; *S. carolinensis* seems to stand this persecution better than does the native red squirrel and therefore tends to spread into unoccupied area from which *S. vulgaris leucourus* has earlier retreated (Shorten, 1953, 1954).

Andrewartha (see also Andrewartha and Birch, 1954) has stressed the apparent fact that while most proponents of the competitive organisation of communities have emphasised competition for food, there is in fact normally more than enough food present. This appears, incidentally, most strikingly in some of Kohn's unpublished data on the genus *Conus*.

The only conclusion that one can draw at present from the observations is that although animal communities appear qualitatively to be constructed as if competition were regulating their structure, even in the best studied cases there are nearly always difficulties and unexplored possibilities. These difficulties suggest that if competition is determinative it either acts intermittently, as in abnormally dry seasons for *Plethodon*, or it is a more subtle process than has been supposed. Thus Lincoln Brower (*in press*) investigating a group of species of North American *Papilio* in which one eastern polyphagous species is replaced by three western oligophagous species, has been impressed by the lack of field evidence for any inadequacy in food resources. He points out however, that specific separation of food might lower the probability of local high density on a given plant, and so the risk of predation by a bird that only stopped to feed when food was abundant (*cf.* de Ruiter, 1952).

Unfortunately there is no end to the possible erection of hypothesis fitted to particular cases that will bring them within the rubric of increasingly subtle forms of competition. Some other method of investigation would clearly be desirable. Before drawing attention to one such possible method, the expected limitations of the Volterra-Gause principle must be examined.

Cases where the Volterra-Gause principle is unlikely to apply. (a) Skellam (1951; see also Brian, 1956b) has considered a model in which two species occur one of (S_1) much lower reproductive potential than the other (S_2). It is assumed that if S_1 and S_2 both arrive in an element of the biotops S_1 always displaces S_2, but that excess elements are always available at the time of breeding and dispersal so that some are never occupied S_1. In view of the higher reproductive potential, S_2 will reach some of these and survive. The model is primarily applicable to annual plants with a definite breeding season, random dispersal of seeds and complete seasonal mortality so all sites are cleared before the new generation starts growing, S_2 is in fact a limiting case of what Hutchinson (1951, 1953) called a fugitive species which could only be established in randomly vacated elements of a biotop. Skellam's model requires clearing of sites by high death rate, Hutchinson's qualitative statement a formation of transient sites by random small catastrophes in the biotop. Otherwise the two concepts developed independently are identical.

(b) When competition for resources becomes a contest rather than a scramble in Nicholson's admirable terminology, there is a theoretical possibility that the principle might not apply. If the breeding population be limited by the number of territories that can be set up in an area, and if a number of unmated individuals without breeding territory are present, food being in excess of the overall requirements, it is possible that territories could be set up by any species entirely independent of the other species, the territorial contests being completely intraspecific. Here a resource, namely area, is limiting but since it does not matter to one species if another is using the area, no interspecific competition need result. No case appears yet to be known, though less extreme modifications of the idea just put forward have apparently been held by several naturalists. Dr. Robert MacArthur has been studying a number of sympatric species of American warblers of the genus *Dendroica* which might be expected to be as likely as any organism to show the phenomenon. He finds however very striking niche specificity among species inhabiting the same trees.

(c) The various cases where circumstances change in the biotop reversing the direction of competition before the latter has run its course. Ideally we may consider two extreme cases with regard to the effect of changing weather and season on competition. In natural populations living for a time under conditions simulating those obtaining in laboratory cultures in a thermostat, if the competition time, that is, the time needed to permit replacement of one species by another, is very short compared with the periods of the significant environmental variables, then complete replacement will occur. This can only happen in very rapidly breeding organisms. Proctor (1957) has found that various green algae always replace *Haematococcus* is small bodies of water which never dry up, though if desiccation and refilling occur frequently enough the *Haematococcus* which is more drought resistant than its competitors will persist indefinitely. If on the contrary the competition time is long compared with the environmental periods, then the relevant environmental determinants of competition will tend to be mean climatic parameters, showing but secular trends in most cases, and competition will inevitably proceed to its end unless some quite exceptional event intervenes.[5]

[5] If there were really three species of giant tortoise (Rothschild, 1915) on Rodriguez, and even more on Mauritius, and if these were sympatric and due to multiple invasion (unlike the races on Albemarle in the Galapagos Islands) it is just conceivable that the population growth was so slow that mixed populations persisted for centuries and that the completion of competition had not occurred before man exterminated all the species involved.

Between the two extreme cases it is reasonable to suppose that there will exist numerous cases in which the direction of competition is never constant enough to allow elimination of one competitor. This seems likely to be the case in the autotrophic plankton of lakes, which inhabits a region in which the supply of nutrients is almost always markedly suboptimal, is subject to continual small changes in temperature and light intensity and in which a large number of species may (Hutchinson, 1941, 1944) coexist.

There is interesting evidence derived from the important work of Brian (1956a) on ants that the completion of competitive exclusion is less likely to occur in seral than in climax stages, which may provide comparable evidence of the effect of environmental changes in competition. Moreover whenever we find the type of situation described so persuasively by Andrewartha and Birch (1954) in which the major limitation on numbers is the length of time that meteorological and other conditions are operating favorably on a species, it is reasonable to suppose that interspecific competition is no more important than intraspecific competition. Much of the apparent extreme difference between the outlook of, for instance, these investigators, or for that matter Milne on the one hand, and a writer such as Lack (1954) on the other, is clearly due to the relationship of generation time to seasonal cycle which differs in the insects and in the birds. The future of animal ecology rests in a realisation not only that different animals have different autecologies, but also that different major groups tend to have fundamental similarities and differences particularly in their broad temporal relationships. The existence of the resemblances moreover may be quite unsuspected and must be determined empirically. In another place (Hutchinson, 1951) I have assembled such evidence as exists on the freshwater copepoda, which seem to be reminiscent of birds rather than of phytoplankton or of terrestrial insects in their competitive relationships.

It is also important to realize, as Cole has indicated in the introductory contribution to this Symposium, that the mere fact that the same species are usually common or rare over long periods of time and that where changes have been observed in well studied faunas such as the British birds or butterflies they can usually be attributed to definite environmental causes in itself indicates that the random action of weather on generation is almost never the whole story. Skellam's demonstration that such action must lead to final extinction must be born in mind. It is quite possible that the change in the phytoplankton of some of the least culturally influenced of the English Lakes, such as the disappearance of *Rhizosolenia* from Wastwater (Pearsall, 1932), may provide a case of random extinction under continually reversing competition. The general evidence of considerable stability under most conditions would suggest that competitive action of some sort is nearly always of significance.

Rarity and commonness of species and the non-intersection of realised niches. Several ways of approaching the problem of the rarity and commonness of species have been suggested (Fisher, Corbet and Williams, 1943; Preston, 1948; Brian, 1953; Shinozaki and Urata, 1953). In all these approaches relatively simple statistical distributions have been fitted to the data, without any attempt being made to elucidate the biological meaning of such distribution. Recently however MacArthur (1957) has advanced the subject by deducing the consequences of certain alternative hypotheses which can be developed in terms of a formal theory of niches.

It has been pointed out in a previous paragraph that the Volterra-Gause principle is equivalent to a statement that the realised niches of co-occurring species are non-interesting. Consider a B space containing an equilibrium community of n species $S_1 S_2 \cdots S_n$, represented by numbers of individuals $N_1 N_2 \cdots N_2$. For any species S_K it will be possible to identify in B a number of elements, each of which corresponds to a whole or part of N'_K and to no other part of N. Suppose that at any given moment each of these elements is occupied by a single individual of S_K, the total volume of B which may be regarded as the specific biotop of S_K will be $N_K \Delta B (S_K)$, $\Delta B(S_K)$ being the mean volume of B occupied by one individual of S_K. Since the biotop is in equilibrium with respect to the n species present, all possible spaces will be filled so that

$$B = \sum_{K=1}^{n} N_K \Delta B (S_K)$$

We do not know anything *a priori* about the distribution of $N_1 \Delta B(S_1), N_2 \Delta B(S_2) \cdots N_n \Delta B(S_n)$, except that these different specific biotops are taken as volumes proportional to $N_1 N_2 \cdots N_n$, which is a justifiable first approximation if the species are of comparable size and physiology. In general some of the species will be rare and some common. The simplest hypothesis consistent with this, is that a random division of B between the species has taken place.

Consider a line of finite length. This may be broken at random into n parts by throwing $(n - 1)$ random points upon it. It would also be possible to divide the line successively by throwing n random pairs of points upon it. In the first case the division is into non-overlapping sections, in the second the sections overlap. MacArthur, whose paper may be consulted for references to the mathematical procedures involved, has given the expected distributions for the division of a line by these alternative methods (Fig. 1). He has moreover shown that with certain restrictions the distribution (I) which corresponds to non-intersecting specific biotops and so to non-intersecting realised niches, fits certain multispecific biological associations extremely well. The form of this distribution is independent of the number of dimensions in B. The alternative distribution with overlapping specific biotops predicts fewer species of intermediate rarity and more of great rarity than is actually found; proceding from the linear case (II), to division of an area or a volume, accentuates this discrepancy. Two very striking cases in which distribution I fits biological multispecific populations are given in Figures 2 and 3 from MacArthur and in Figure 4 from the recent studies of Dr. Alan Kohn (in press).

The limitation which is imposed by the theory is that in all large subdivisions of B the ratio of total number of individuals ($m = \sum_{i=1}^{n} N_i$) to

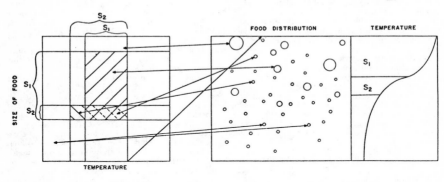

FIGURE 1. Two fundamental niches defined by a pair of variables in a two-dimensional niche space. Only one species is supposed to be able to persist in the intersection subset region. The lines joining equivalent points in the niche space and biotop space indicate the relationship of the two spaces. The distribution of the two species involved is shown on the right hand panel with a temperature depth curve of the kind usual in a lake in summer.

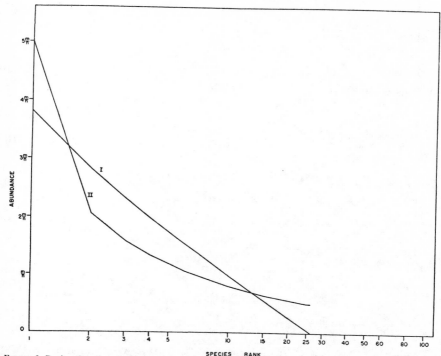

FIGURE 2. Rank order of species arranged per number of individuals according to the distributions I and II considered by MacArthur.

total number of species (n) must remain constant. This is likely to be the case in any biotop which is what may be termed *homogeneously diverse*, that is, in which the elements of the environmental mosaic (trees, stones, bushes, dead logs, etc.) are small compared with the mean free paths of the organisms under consideration. When a heterogeneously diverse area, comprising for instance stands of woodland separated by areas of pasture, is considered it is very unlikely that the ratio of total numbers of individuals to number of species will be identical in both woodland and pasture (if it occasionally were, the fact that both censuses could be added would not be of any biological interest). MacArthur finds that at least some bodies of published data which do not fit distribution I as a whole, can be broken down according to the type of environment into subcensuses which do fit the distribution. Data from moth traps and from populations of diatoms on slides submerged in rivers would not be expected to fit the distribution and in fact do not do so.[6] Such collection methods certainly sample very heterogeneously diverse areas.

The great merit of MacArthur's study is that it attempts to deduce operationally distinct differences between the results of two rival hypotheses, one of which corresponds essentially to the extreme density dependent view of interspecific interaction, the other to the opposite view. Although certain simplifying assumptions must be made in the theoretical treatment, the initial results suggest that in stable homogeneously diverse biotops the abundances of different species are arranged as if the realised niches were non-overlapping; this does not mean that populations may not exist under other conditions which would depart very widely from MacArthur's findings.

The problem of the saturation of the biotop. An important but quite inadequately studied aspect

[6] I am indebted to Dr. Ruth Patrick for the opportunity to test some of her diatometer censuses.

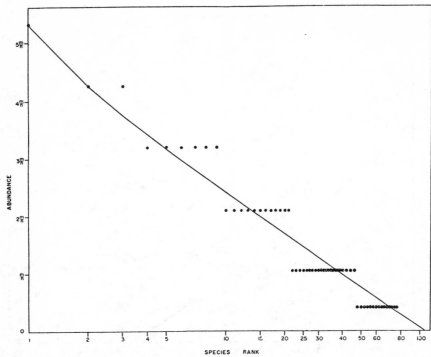

FIGURE 3. Rank order of species of birds in a tropical forest, closely following MacArthur's distribution I.

of niche specificity is that of the number of species that a given biotop can support. The nature of this problem can be best made clear by means of an example.

The aquatic bugs of the family *Corixidae* are of practically world wide distribution. Omitting a purely Australasian subfamily, they may be divided into the *Micronectinae* which are nearly always small, under 5 mm long and the *Corixinae* of which the great majority of species are over 5 mm long. Both subfamilies probably feed largely on organic detritus, though a few of the more primitive members of the *Corixinae* are definite predators. Some at least suck out the contents of algal cells, but unlike the other Heteroptera they can take particulate matter of some size unto their alimentary tracts. There is abundant evidence that the organic content of the bottom deposits of the shallow water in which these insects live is a major ecological factor regulating their occurrence. No *Micronectinae* occur in temperate North America and in the Old World this subfamily is far more abundant in the tropics while the *Corixnae* are far more abundant in the temperate regions (Lundblad,1934;Jaczewski,1937).Thus in Britain there are 30 species of *Corixinae* and three of *Micronectinae* (Macan, 1956), in peninsular Italy 20 or 21 species of *Corixinae* and five of *Micronectinae* (Stickel, 1955), in non-Palaeartic India about a dozen species of *Corixinae* and at least ten species of *Micronectinae* (Hutchinson, 1940) and in Indonesia (Lundblad, 1934) only three *Corixinae* and 14 *Micronectinae*. A reasonable explanation of this variation in the relative proportions of the two subfamilies is suggested by the findings of Macan (1938) and the more casual observations of other investigators that *Micronecta* prefers a low organic subtratum; in tropical localities the high rate of decomposition would reduce the organic content.

In certain isolated tropical areas at high altitudes, notably Ethiopia and the Nilghiri Hills of southern India the decline in the numbers of *Micronectinae* with increasing altitudes, and so

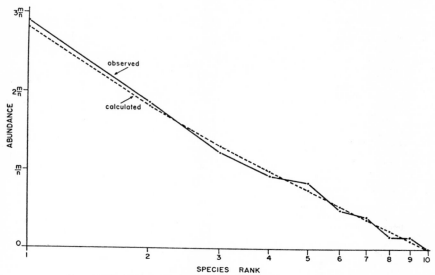

FIGURE 4. Rank order of species of *Conus* on a littoral bench in Hawaii (Kohn).

lower average water temperatures, is most noticable, but there is no increase in the number of Corixinae, presumably because the surrounding fauna is not rich enough to have permitted frequent invasion and speciation. Thus in the Nilghiri Hills between 2100 and 2300 m, intense collecting yielded three *Corixinae* of which two appear to be endemic, and one non-endemic species of *Micronecta*. Very casual collecting below 1000 m in south India has produced two species of *Corixinae* and five species of *Micronectinae*. The question raised by cases like this is whether the three Nilghiri *Corixinae* fill all the available niches which in Europe might support perhaps 15 or 20 species, or whether there are really empty niches. Intuitively one would suppose both alternatives might be partly true, but there is no information on which to form a real judgment. The rapid spread of introduced species often gives evidence of empty niches, but such rapid spread in many instances has taken place in disturbed areas. The problem clearly needs far more systematic study than it has been given. The addition and the replacement of species of fishes proceeding down a river, and the competitive situations involved, may provide some of the best material for this sort of study, but though much data exists, few attempts at systematic comparative interpretation have been made (*cf.* Hutchinson, 1939).

THE FUTURE OF COMPARATIVE DEMOGRAPHIC STUDIES

Perhaps the most interesting general aspect of the present Symposium is the strong emphasis placed on the changing nature of the populations with which almost all investigators deal. In certain cases, notably in the parthenogenetic crustacean *Daphnia* (Slobodkin, 1954), it is possible to work with clones that must be almost uniform genetically, but all the work on bisexual organisms is done under conditions in which evolution may take place. The emergence in Nicholson's experiments of strains of *Lucilia* in which adult females no longer need a protein meal before egg laying provides a dramatic example of evolution in the laboratory; the work reported by Dobzhansky, by Lewontin, and by Wallace, in discussion, shows how experimental evolution, for which subject the Carnegie Laboratory at Cold Spring Harbor was founded, has at last come into its own.

So far little attention has been paid to the problem of changes in the properties of populations of the greatest demographic interest in such experiments. A more systematic study of evolutionary change in fecundity, mean life span, age and duration of reproductive activity and length of post reproductive life is clearly needed. The most interesting models that might be devised would be those in which selection operated in favor of low fecundity, long pre-reproductive life and on any aspect of post-reproductive life.

There is in many groups, notably *Daphnia*, dependence of natality on food supply (Slobodkin, 1954) though the adjustment can never be instantaneous and so can lead to oscillations. In the case of birds the work of the Oxford school (Moreau, 1944; Lack, 1947a, 1954 and many papers quoted in the last named) indicates that in many birds natality is regulated by natural selection to correspond to the maximum number of young that can be reared in a clutch. In some circumstances the absolute survival of young is greater when the fecundity is low than when it is high. The peculiar nature of the subpopulations formed by groups of nestlings in nests makes this reasonable. Slobodkin (1953) has pointed out that in certain cases in which migration into numerous limited areas is possible, a high reproductive rate might have a lower selective advantage than a low rate. Actually in a very broad sense the bird's nest is a device to formalise the numerous limited areas, the existence of which permits such a type of selection. It should be possible with some insects to set up population cages in which access to a large number of very small amounts of larval food is fairly difficult for a fertile female. If the individual masses of larval food were such that there was an appreciable chance that many larvae on a single mass would die of starvation while a few larvae would survive, it is possible that selection for low fecundity might occur. This experiment would certainly imitate many situations in nature.

The evolutionary aspects of the problem raised by those cases where there is a delay of reproductive activity after adult morphology has been achieved is much harder to understand. Some birds though they attain full body size within a year (or in the case of most passerines in the nest) are apparently not able to breed until their third or later year. It is difficult to see why this should be so. In any given species there may be good endocrinological reasons for the delay, but they can hardly be evolutionarily inevitable. The situation has an obvious *prima facie* disadvantage, since most birds have a strikingly diagonal survivorship curve after the first year of life and this in itself indicates little capacity for learning to live. One would have supposed that in the birds, mainly but not exclusively large sea birds, which show the delay, any genetic change favoring early reproduction would have a great selective advantage. Any experimental model imitating this situation would be of great interest.

The problem of possible social effects of long post-reproductive life, which can hardly be subject to direct selection, provides another case in which any hints from changes in demographic parameters in experiments would be most helpful. The experimental study of the evolutionary aspects of demography is certain to yield surprises. While we have Nicholson's work, in which the amplitude of the oscillation in *Lucilia* populations appear to be increased or at least not decreased as a result of the evolution he has observed, though the minima are less low and the variation less regular, we do not know if this sort of effect is likely to be general. Utida's elegant work on bean weevils appears to be consistent with some evolutionary damping of oscillations which would be theoretically a likely result.

The most curious case of a genetic change playing a regular part in a demographic process is certainly that in rodents described by Chitty. In view of the large number of simple ways which are now available to explain regular oscillations in a population, it is extremely important to heed Chitty's warning that the obvious explanation is not necessarily the true one. To the writer, this seems to be a particular danger in human demography, though the mysteries of variation of the human sex ratio, so clearly expounded by Colombo, should be a warning against over-simple hypotheses, for here no reasonable hypotheses have been suggested. Human demography relies too much on what psychologists call intervening variable theory. The reproducing organisms are taken for granted; when their properties change, either as the result of evolution or of changes in learned behaviour, the results are apt to be upsetting. The present "baby boom" is such an upset, and here a tendency to over-simplified thinking is also apparent. If, as appears clear at least for parts of North America, the present birth rate is positively correlated with economic position, it is easy to suppose that couples now have as many children as they can afford, just as most small birds appear to do. There is, however, a difference. If at any economic level a four child family was desired, but occasionally owing to the imperfections of birth control a five child family was actually achieved, we should not expect the fifth child to have a negligible expectation of life at birth, so that the total contribution to the population per family would be the same from a four and a five child family. Yet this is exactly what Lack and Arn (1947) found for the broods of the Alpine swift *Apus melba*. In man the criterion is never purely economic; it is not how large a brood can be reared, but how large a brood the parents think they can rear without undue economic sacrifice. Such a method of setting limits to natality is obviously extremely complicated. It involves an equilibrium between a series of desires, partly conscious, partly unconscious, and a series of estimates of present and future resources. There is absolutely no reason to suppose that the mean desired family size determined in such a way is a simple function of economics, uninfluenced by a vast number of other cultural factors. The assumption that a large family is *per se* a good thing is obviously involved; this may be accepted individually by most parents even though it is at present a very dubious assumption on general grounds of social well being. Part of the acceptance of such an assumption is certain to be due to unconscious factors. Susannah Coolidge in a remarkable, as yet unpublished, essay,[7] "Population versus People," suggests that for many women a new pregnancy is an occasion for a temporary shifting of some of the responsibility for the older children away from the mother, and so is welcomed. She also suspects that it may be an unconscious expression of disappointment over, or repudiation of, the older children and so be essentially a repeated neurotic symptom. Moreover, the present highly conspicuous fashion for maternity, certainly a healthy reaction from the seclusion of upper-class pregnant women a couple of generations ago, is also quite likely fostered by those business interests which seem to believe that an indefinitely expanding economy is possible on a non-expanding planet.

An adequate science of human demography must take into account mechanism of these kinds, just as animal demography has taken into account all the available information on the physiological ecology and behaviour of blow flies, *Daphnia* and bean weevils. Unhappily, human beings are far harder to investigate than are these admirable laboratory animals; unhappily also, the need becomes more urgent daily.

REFERENCES

ANDREWARTHA, H. G., and BIRCH, L. C., 1954, The Distribution and Abundance of Animals. Chicago, University of Chicago Press. xv, 782 pp.
BETTS, M. M., 1955, The food of titmice in oak woodland. J. Anim. Ecol. *24:* 282–323.
BIRCH, L. C., 1957, The meanings of competition. Amer. Nat. *91:* 5–18.
BRIAN, M. V., 1953, Species frequencies from random samples in animal populations. J. Anim. Ecol. *22:* 57–64.
1956a, Segregation of species of the ant genus *Myrmica*. J. Anim. Ecol. *25:* 319–337.
1956b, Exploitation and interference in interspecies competition. J. Anim. Ecol. *25:* 339–347.
COOPER, D. M., and DOBZHANSKY, TH., 1956, Studies on the ecology of *Drosophila* in the Yosemite region of California. I. The occurrence of species of *Drosophila* in different life zones and at different seasons. Ecology *37:* 526–533.
CROMBIE, A. C., 1945, On competition between different species of graminivorous insects. Proc. Roy. Soc. Lond. *132*B: 362–395.
1946, Further experiments on insect competition. Proc. Roy. Soc. Lond. *133*B: 76–109.
1947, Interspecific competition. J. Anim. Ecol. *16:* 44–73.
DA CUNHA, A. B., EL-TABEY SHEKATA, A. M., and DE OLIVIERA, W., 1957, A study of the diet and nutritional preferences of tropical of *Drosophila*. Ecology *38:* 98–106.
DUMAS, P. C., 1956, The ecological relations of sympatry in *Plethodon dunni* and *Plethodon vehiculum*. Ecology *37:* 484–495.
FISHER, R. A., CORBET, A. S., and WILLIAMS, C. B., 1943, The relation between the number of species and the number of individuals in a ransom sample of an animal population. J. Anim. Ecol. *12:* 42–58.
GAUSE, G. F., 1934, The struggle for existence. Baltimore, Williams & Wilkins. 163 pp.
1935, Vérifications expérimentales de la théorie mathématique de la lutte pour la vie. Actualités scientifiques *277*. Paris. 63 pp.
HUTCHINSON, G. E., 1939, Ecological observations on the fishes of Kashmir and Indian Tibet. Ecol. Monogr. *9:* 145–182.
1940, A revision of the Corixidae of India and adjacent regions. Trans. Conn. Acad. Arts Sci. *33:* 339–476.
1941, Ecological aspects of succession in natural populations. Amer. Nat. *75:* 406–418.
1944, Limnological studies in Connecticut. VII. A critical examination of the supposed relationship between phytoplankton periodicity and chemical changes in lake waters. Ecology *25:* 3–26.
1951, Copepodology for the ornithologist. Ecology *32:* 571–577.
1953, The concept of pattern in ecology. Proc. Acad. Nat. Sci. Phila. *105:* 1–12.
HYNES, H. B. N., 1954, The ecology of *Gammarus deubeni* Lilljeborg and its occurrence in fresh water in western Britain. J. Anim. Ecol. *23:* 38–84.
1955, The reproductive cycle of some British freshwater Gammaridae. J. Anim. Ecol. *24:* 352–387.
JACZEWSKI, S., 1937, Allgemeine Zügeder geographischen Verbreitung der Wasserhemiptera. Arch. Hydrobiol. *31:* 565–591.
KOHN, A. J., The ecology of *Conus* in Hawaii. (Yale Dissertation 1057, in press.)
KURTÉN, B., 1957, Mammal migrations, cenozoic stratigraphy, and the age of Peking man and the australopithecines. J. Paleontol. *31:* 215–227.
LACK, D., 1947a, The significance of clutch-size. Ibis *89:* 302–352.
1947b, Darwin's finches. Cambridge, England. x, 208 pp.
1954, The natural regulation of animal numbers. Oxford, The Clarendon Press, viii, 343.
LACK, D., and ARN, H., 1947, Die Bedeutung der Gelegegrösse beim Alpensegler. Ornith. Beobact. *44:* 188–210.
LOTKA, A. J., The growth of mixed populations, two species competing for a common food supply. J. Wash. Acad. Sci. *22:* 461–469.
LUNDBLAD, O., 1933, Zur Kenntnis der aquatilen und semi-aquatilen Hemipteren von Sumatra, Java, und Bali. Arch. Hydrobiol. Suppl. *12:* 1–195, 263–489.
MACAN, T. T., 1938, Evolution of aquatic habitats with special reference to the distribution of Corixidae. J. Anim. Ecol. *7:* 1–19.
1956, A revised key to the British water bugs (Hemiptera, Heteroptera) Freshwater Biol. Assoc. Sci. Publ. *16:* 73 pp.
MACARTHUR, R. H., 1957, On the relative abundance of bird species. Proc. Nat. Acad. Sci. Wash. *43:* 293–295.
MOREAU, R. E., 1944, Clutch-size: a comparative study, with special reference to African birds. Ibis *86:* 286–347.
PEARSALL, W. H., 1932, The phytoplankton in the English lakes II. The composition of the phytoplankton in relation to dissolved substances. J. Ecol. *20:* 241–262.
PRESTON, F. W., 1948, The commonness, and rarity, of species. Ecology *29:* 254–283.
PROCTOR, V. W., 1957, Some factors controlling the distribution of *Haematococcus pluvialis*. Ecology, in press.
ROTHSCHILD, LORD, 1915, On the gigantic land tortoises of the Seychelles and Aldabra-Madagascar group with some notes on certain forms of the Mascarene group. Novitat. Zool. *22:* 418–442.
RUITER, L. DE, 1952, Some experiments on the camouflage of stick caterpillars. Behaviour *4:* 222–233.

---

[7] I am greatly indebted to the author of this work for permission to refer to some of her conclusions.

SHINOZAKI, K., and URATA, N., 1953, Researches on population ecology II. Kyoto Univ. (not seen; ref. MacArthur, 1957).

SHORTEN, M., 1953, Notes on the distribution of the grey squirrel (*Sciurus carolinensis*) and the red squirrel (*Sciurus vulgaris leucourus*) in England and Wales from 1945 to 1952. J. Anim. Ecol. *22:* 134–140.

—— 1954, Squirrels. (New Naturalist Monograph 12.) London 212 pp.

SKELLAM, J. G., 1951, Random dispersal in theoretical populations. Biometrika *38:* 196–218.

SLOBODKIN, L. B., 1953, An algebra of population growth. Ecology *34:* 513–517.

—— 1954, Population dynamics in *Daphnia obtusa* Kurz. Ecol. Monogr. *24:* 69–88.

STEUSLOFF, V., 1943, Ein Beitrag zur Kenntniss der Verbreitung und der Lebensräume von *Gammarus*-Arten in Nordwest-Deutschland. Arch. Hydrobiol. *40:* 79–97.

STICKEL, W., 1955, Illustrierte Bestimmungstabellen der Wanzen. II. Europa. Hf. 2, 3. pp. 40–80 (Berlin, apparently published by author).

UENO, M., 1934, Yale North India Expedition. Report on the amphipod genus *Gammarus*. Mem. Conn. Acad. Arts Sci. *10:* 63–75.

VOLTERRA, V., 1926, Vartazioni e fluttuazioni del numero d'individui in specie animali conviventi. Mem. R. Accad. Lincei ser. 6, *2:* 1–36.

WINSOR, C. P., 1934, Mathematical analysis of the growth of mixed populations. Cold Spr. Harb. Symp. Quant. Biol. *2:* 181–187.

NICHE SEGREGATION IN SEVEN SPECIES OF DIPLOPODS
Robert V. O'Neill

The concept of competitive exclusion has been severely criticized (Cole 1960), but evidence in its favor continues to be produced (Dybas and Lloyd 1962, Caldwell and Gentry 1965, McCarley 1963). It is frequently the case that superficial examination shows sympatric populations that appear to be in direct competition. Such a case is found with seven genera of millipedes in maple-oak forests in Central Illinois. All are saprophagous, feeding on leaf litter or decaying wood and all would appear to be occupying the same niche. More intensive investigation has revealed, however, that the populations are separated by subtle but real differences in microhabitat. Table 1 records the exact physical location of each millipede seen over a period from 1964 to 1966. When the habitat is divided into seven distinct microhabitats, it can be seen that each species occupies one predominantly, and that each microhabitat is occupied predominantly by only one species. It appears that the millipedes have divided the habitat spatially into subunits within which competition is reduced.

Literature Cited

Caldwell, L. D. and J. B. Gentry. 1965. Interactions of *Peromyscus* and *Mus* in a one-acre field enclosure. Ecology **46**: 189–192.

Cole, L. C. 1960. Competitive Exclusion. Science **132**: 348–349.

Dybas, H. S. and M. Lloyd. 1962. Isolation by habitat in two synchronized species of periodical cicadas. Ecology **43**: 444–449.

McCarley, H. 1963. Distributional relationships of sympatric populations of *Peromyscus leucopus* and *P. gossypinus*. Ecology **44**: 784–788.

ECOLOGY, 1967, Vol. 48, p. 983.

TABLE 1. Percentages of each species found in seven microhabitats

Habitats	Euryurus erythropygus (Brandt) (133)[1]	Pseudopoly-desmus serratus (Say) (50)	Narceus americanus (Beauvois) (231)	Scytonotus granulatus (Say) (35)	Fontaria virginiensis (Drury) (135)	Cleidogonia caesioannularis (Wood) (25)	Abacion lacterium (Say) (50)
Heartwood at center of logs	93.9[2]	0	0	0	0	0	0
Superficial wood of logs	0	66.7	4.3	6.7	0	14.3	0
Outer surface of logs beneath bark	0	20.8	71.4	0	0	0	15.8
Under log, but on log surface	3.0	8.3	6.9	60.0	0	0	36.8
Under log, but on ground surface	3.0	4.2	12.5	0	97.1	14.3	0
Within leaves of litter	0	0	0	26.7	0	42.8	0
Beneath litter on ground surface	0	0	4.7	6.7	2.9	28.6	47.4

[1] Total number of individuals observed
[2] Underlined figures represent microhabitats in which the species predominate.

THE INFLUENCE OF INTERSPECIFIC COMPETITION AND OTHER FACTORS ON THE DISTRIBUTION OF THE BARNACLE *CHTHAMALUS STELLATUS*

JOSEPH H. CONNELL

INTRODUCTION

Most of the evidence for the occurrence of interspecific competition in animals has been gained from laboratory populations. Because of the small amount of direct evidence for its occurrence in nature, competition has sometimes been assigned a minor role in determining the composition of animal communities.

Indirect evidence exists, however, which suggests that competition may sometimes be responsible for the distribution of animals in nature. The range of distribution of a species may be decreased in the presence of another species with similar requirements (Beauchamp and Ullyott 1932, Endean, Kenny and Stephenson 1956). Uniform distribution is space is usually attributed to intraspecies competition (Holme 1950, Clark and Evans 1954). When animals with similar requirements, such as 2 or more closely related species, are found coexisting in the same area, careful analysis usually indicates that they are not actually competing with each other (Lack 1954, MacArthur 1958).

In the course of an investigation of the animals of an intertidal rocky shore I noticed that the adults of 2 species of barnacles occupied 2 separate horizontal zones with a small area of overlap, whereas the young of the species from the upper zone were found in much of the lower zone. The upper species, *Chthamalus stellatus* (Poli) thus settled but did not survive in the lower zone. It seemed probable that this species was eliminated by the lower one, *Balanus balanoides* (L), in a struggle for a common requisite which was in short supply. In the rocky intertidal region, space for attachment and growth is often extremely limited. This paper is an account of some observations and experiments designed to test the hypothesis that the absence in the lower zone of adults of *Chthamalus* was due to interspecific competition with *Balanus* for space. Other factors which may have influenced the distribution were also studied. The study was made at Millport, Isle of Cumbrae, Scotland.

I would like to thank Prof. C. M. Yonge and the staff of the Marine Station, Millport, for their help, discussions and encouragement during the course of this work. Thanks are due to the following for their critical reading of the manuscript: C. S. Elton, P. W. Frank, G. Hardin, N. G. Hairston, E. Orias, T. Park and his students, and my wife.

Distribution of the species of barnacles

The upper species, *Chthamalus stellatus*, has its center of distribution in the Mediterranean; it reaches its northern limit in the Shetland Islands, north of Scotland. At Millport, adults of this species occur between the levels of mean high water of neap and spring tides (M.H.W.N. and M.H.W.S.: see Figure 5 and Table I). In southwest England and Ireland, adult *Chtham-*

alus occur at moderate population densities throughout the intertidal zone, more abundantly when *Balanus balanoides* is sparse or absent (Southward and Crisp 1954, 1956). At Millport the larvae settle from the plankton onto the shore mainly in September and October; some additional settlement may occur until December. The settlement is most abundant between M.H.W.S. and mean tide level (M.T.L.), in patches of rock surface left bare as a result of the mortality of *Balanus*, limpets, and other sedentary organisms. Few of the *Chthamalus* that settle below M.H.W.N. survive, so that adults are found only occasionally at these levels.

Balanus balanoides is a boreal-arctic species, reaching its southern limit in northern Spain. At Millport it occupies almost the entire intertidal region, from mean low water of spring tides (M.L.W.S.) up to the region between M.H.W.N. and M.H.W.S. Above M.H.W.N. it occurs intermingled with *Chthamalus* for a short distance. *Balanus* settles on the shore in April and May, often in very dense concentrations (see Table IV).

The main purpose of this study was to determine the cause of death of those *Chthamalus* that settled below M.H.W.N. A study which was being carried on at this time had revealed that physical conditions, competition for space, and predation by the snail *Thais lapillus* L. were among the most important causes of mortality of *Balanus balanoides*. Therefore, the observations and experiments in the present study were designed to detect the effects of these factors on the survival of *Chthamalus*.

METHODS

Intertidal barnacles are very nearly ideal for the study of survival under natural conditions. Their sessile habit allows direct observation of the survival of individuals in a group whose positions have been mapped. Their small size and dense concentrations on rocks exposed at intervals make experimentation feasible. In addition, they may be handled and transplanted without injury on pieces of rock, since their opercular plates remain closed when exposed to air.

The experimental area was located on the Isle of Cumbrae in the Firth of Clyde, Scotland. Farland Point, where the study was made, comprises the southeast tip of the island; it is exposed to moderate wave action. The shore rock consists mainly of old red sandstone, arranged in a series of ridges, from 2 to 6 ft high, oriented at right angles to the shoreline. A more detailed description is given by Connell (1961). The other barnacle species present were *Balanus crenatus* Brug and *Verruca stroemia* (O. F. Muller), both found in small numbers only at and below M.L.W.S.

To measure the survival of *Chthamalus*, the positions of all individuals in a patch were mapped. Any barnacles which were empty or missing at the next examination of this patch must have died in the interval, since emigration is impossible. The mapping was done by placing thin glass plates (lantern slide cover glasses, 10.7×8.2 cm, area 87.7 cm^2) over a patch of barnacles and marking the position of each *Chthamalus* on it with glass-marking ink. The positions of the corners of the plate were marked by drilling small holes in the rock. Observations made in subsequent censuses were noted on a paper copy of the glass map.

The study areas were chosen by searching for patches of *Chthamalus* below M.H.W.N. in a stretch of shore about 50 ft long. When 8 patches had been found, no more were looked for. The only basis for rejection of an area in this search was that it contained fewer than 50 *Chthamalus* in an area of about 1/10 m^2. Each numbered area consisted of one or more glass maps located in the 1/10 m^2. They were mapped in March and April, 1954, before the main settlement of *Balanus* began in late April.

Very few *Chthamalus* were found to have settled below mid-tide level. Therefore pieces of rock bearing *Chthamalus* were removed from levels above M.H.W.N. and transplanted to and below M.T.L. A hole was drilled through each piece; it was then fastened to the rock by a stainless steel screw driven into a plastic screw anchor fitted into a hole drilled into the rock. A hole ¼" in diameter and 1" deep was found to be satisfactory. The screw could be removed and replaced repeatedly and only one stone was lost in the entire period.

For censusing, the stones were removed during a low tide period, brought to the laboratory for examination, and returned before the tide rose again. The locations and arrangements of each area are given in Table I; the transplanted stones are represented by areas 11 to 15.

The effect of competition for space on the survival of *Chthamalus* was studied in the following manner: After the settlement of *Balanus* had stopped in early June, having reached densities of 49/cm^2 on the experimental areas (Table I) a census of the surviving *Chthamalus* was made on each area (see Figure 1). Each map was then divided so that about half of the number of

TABLE I. Description of experimental areas*

Area no.	Height in ft from M.T.L.	% of time submerged	Population Density: no./cm² in June, 1954			Remarks
			Chthamalus, autumn 1953 settlement		All barnacles, undisturbed portion	
			Undisturbed portion	Portion without Balanus		
MHWS	+4.9	4	—	—	—	—
1	+4.2	9	2.2	—	19.2	Vertical, partly protected
2	+3.5	16	5.2	4.2	—	Vertical, wave beaten
MHWN	+3.1	21	—	—	—	—
3a	+2.2	30	0.6	0.6	30.9	Horizontal, wave beaten
3b	"	"	0.5	0.7	29.2	" " " "
4	+1.4	38	1.9	0.6	—	30° to vertical, partly protected
5	+1.4	"	2.4	1.2	—	" " " " " "
6	+1.0	42	1.1	1.9	38.2	Horizontal, top of a boulder, partly protected
7a	+0.7	44	1.3	2.0	49.3	Vertical, protected
7b	"	"	2.3	2.0	51.7	" "
11a	0.0	50	1.0	0.6	32.0	Vertical, protected
11b	"	"	0.2	0.3	—	" "
12a	0.0	100	1.2	1.2	18.8	Horizontal, immersed in tide pool
12b	"	100	0.8	0.9	—	" " " " " "
13a	−1.0	58	4.9	4.1	29.5	Vertical, wave beaten
13b	"	"	3.1	2.4	—	" " "
14a	−2.5	71	0.7	1.1	—	45° angle, wave beaten
14b	"	"	1.0	1.0	—	" " " "
MLWN	−3.0	77	—	—	—	—
MLWS	−5.1	96	—	—	—	—
15	+1.0	42	32.0	—	—	Chthamalus of autumn, 1954 settlement; densities of Oct., 1954.
7b	+0.7	44	5.5	3.7	—	

* The letter "a" following an area number indicates that this area was enclosed by a cage; "b" refers to a closely adjacent area which was not enclosed. All areas faced either east or south except 7a and 7b, which faced north.

Chthamalus were in each portion. One portion was chosen (by flipping a coin), and those Balanus which were touching or immediately surrounding each Chthamalus were carefully removed with a needle; the other portion was left untouched. In this way it was possible to measure the effect on the survival of Chthamalus both of intraspecific competition alone and of competition with Balanus. It was not possible to have the numbers or population densities of Chthamalus exactly equal on the 2 portions of each area. This was due to the fact that, since Chthamalus often occurred in groups, the Balanus had to be removed from around all the members of a group to ensure that no crowding by Balanus occurred. The densities of Chthamalus were very low, however, so that the slight differences in density between the 2 portions of each area can probably be disregarded; intraspecific crowding was very seldom observed. Censuses of the Chthamalus were made at intervals of 4-6 weeks during the next year; notes were made at each census of factors such as crowding, undercutting or smothering which had taken place since the last examination. When necessary, Balanus which had grown until they threatened to touch the Chthamalus were removed in later examinations.

To study the effects of different degrees of immersion, the areas were located throughout the tidal range, either in situ or on transplanted stones, as shown in Table I. Area 1 had been under observation for 1½ years previously. The effects of different degrees of wave shock could not be studied adequately in such a small area

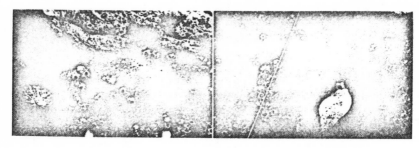

APRIL 16, 1954　　　　　　　　　　JUNE 11, 1954

NOV. 3, 1954　　　　　　　　　　MAY 13, 1955

FIG. 1. Area 7b. In the first photograph the large barnacles are *Balanus*, the small ones scattered in the bare patch, *Chthamalus*. The white line on the second photograph divides the undisturbed portion (right) from the portion from which *Balanus* were removed (left). A limpet, *Patella vulgata*, occurs on the left, and predatory snails, *Thais lapillus*, are visible.

of shore but such differences as existed are listed in Table I.

The effects of the predatory snail, *Thais lapillus*, (synonymous with *Nucella* or *Purpura*, Clench 1947), were studied as follows: Cages of stainless steel wire netting, 8 meshes per inch, were attached over some of the areas. This mesh has an open area of 60% and previous work (Connell 1961) had shown that it did not inhibit growth or survival of the barnacles. The cages were about 4 × 6 inches, the roof was about an inch above the barnacles and the sides were fitted to the irregularities of the rock. They were held in place in the same manner as the transplanted stones. The transplanted stones were attached in pairs, one of each pair being enclosed in a cage (Table I).

These cages were effective in excluding all but the smallest *Thais*. Occasionally small *Thais*, ½ to 1 cm in length, entered the cages through gaps at the line of juncture of netting and rock surface. In the concurrent study of *Balanus* (Connell 1961), small *Thais* were estimated to have occurred inside the cages about 3% of the time.

All the areas and stones were established before the settlement of *Balanus* began in late April, 1954. Thus the *Chthamalus* which had settled naturally on the shore were then of the 1953 year class and all about 7 months old. Some *Chthamalus* which settled in the autumn of 1954 were followed until the study was ended in June, 1955. In addition some adults which, judging from their large size and the great erosion of their shells, must have settled in 1952 or earlier, were present on the transplanted stones. Thus records were made of at least 3 year-classes of *Chthamalus*.

RESULTS

The effects of physical factors

In Figures 2 and 3, the dashed line indicates the survival of *Chthamalus* growing without contact with *Balanus*. The suffix "a" indicates that the area was protected from *Thais* by a cage.

In the absence of *Balanus* and *Thais*, and protected by the cages from damage by water-borne objects, the survival of *Chthamalus* was good at all levels. For those which had settled normally on the shore (Fig. 2), the poorest survival was on the lowest area, 7a. On the transplanted stones (Fig. 3, area 12), constant immersion in a tide pool resulted in the poorest survival. The reasons for the trend toward slightly greater mortality as the degree of immersion increased are unknown. The amount of attached algae on the stones in the tide pool was much greater than on the other areas. This may have reduced the flow of water and food or have interfered directly with feeding movements. Another possible indirect effect of increased immersion is the increase in predation by the snail, *Thais lapillus*, at lower levels.

Chthamalus is tolerant of a much greater degree of immersion than it normally encounters. This is shown by the survival for a year on area 12 in a tide pool, together with the findings of Fischer (1928) and Barnes (1956a), who found that *Chthamalus* withstood submersion for 12 and 22 months, respectively. Its absence below M.T.L. can probably be ascribed either to a lack of initial settlement or to poor survival of newly settled larvae. Lewis and Powell (1960) have suggested that the survival of *Chthamalus* may be favored by increased light or warmth during emersion in its early life on the shore. These conditions would tend to occur higher on the shore in Scotland than in southern England.

The effects of wave action on the survival of *Chthamalus* are difficult to assess. Like the degree of immersion, the effects of wave action may act indirectly. The areas 7 and 12, where relatively poor survival was found, were also the areas of least wave action. Although *Chthamalus* is usually abundant on wave beaten areas and absent from sheltered bays in Scotland, Lewis and Powell (1960) have shown that in certain sheltered bays it may be very abundant. Hatton (1938) found that in northern France, settlement and growth rates were greater in wave-beaten areas at M.T.L., but, at M.H.W.N., greater in sheltered areas.

At the upper shore margins of distribution *Chthamalus* evidently can exist higher than *Balanus* mainly as a result of its greater tolerance to heat and/or desiccation. The evidence for this was gained during the spring of 1955. Records from a tide and wave guage operating at this time about one-half mile north of the study area showed that a period of neap tides had coincided with an unusual period of warm calm weather in April so that for several days no water, not even waves, reached the level of Area 1. In the period

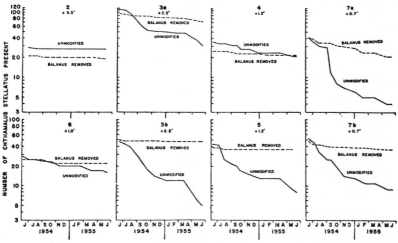

FIG. 2. Survivorship curves of *Chthamalus stellatus* which had settled naturally on the shore in the autumn of 1953. Areas designated "a" were protected from predation by cages. In each area the survival of *Chthamalus* growing without contact with *Balanus* is compared to that in the undisturbed area. For each area the vertical distance in feet from M.T.L. is shown.

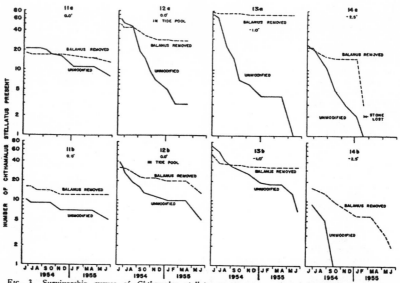

FIG. 3. Survivorship curves of *Chthamalus stellatus* on stones transplanted from high levels. These had settled in the autumn of 1953; the arrangement is the same as that of Figure 2.

between the censuses of February and May, *Balanus* aged one year suffered a mortality of 92%, those 2 years and older, 51%. Over the same period the mortality of *Chthamalus* aged 7 months was 62%, those 1½ years and older, 2%. Records of the survival of *Balanus* at several levels below this showed that only those *Balanus* in the top quarter of the intertidal region suffered high mortality during this time (Connell 1961).

Competition for space

At each census notes were made for individual barnacles of any crowding which had occurred since the last census. Thus when one barnacle started to grow up over another this fact was noted and at the next census 4-6 weeks later the progress of this process was noted. In this way a detailed description was built up of these gradually occurring events.

Intraspecific competition leading to mortality in *Chthamalus* was a rare event. For areas 2 to 7, on the portions from which *Balanus* had been removed, 167 deaths were recorded in a year. Of these, only 6 could be ascribed to crowding between individuals of *Chthamalus*. On the undisturbed portions no such crowding was observed. This accords with Hatton's (1938) observation that he never saw crowding between individuals of *Chthamalus* as contrasted to its frequent occurrence between individuals of *Balanus*.

Interspecific competition between *Balanus* and *Chthamalus* was, on the other hand, a most important cause of death of *Chthamalus*. This is shown both by the direct observations of the process of crowding at each census and by the differences between the survival curves of *Chthamalus* with and without *Balanus*. From the periodic observations it was noted that after the first month on the undisturbed portions of areas 3 to 7 about 10% of the *Chthamalus* were being covered as *Balanus* grew over them; about 3% were being undercut and lifted by growing *Balanus*; a few had died without crowding. By the end of the 2nd month about 20% of the *Chthamalus* were either wholly or partly covered by *Balanus*; about 4% had been undercut; others were surrounded by tall *Balanus*. These processes continued at a lower rate in the autumn and almost ceased during the later winter. In the spring *Balanus* resumed growth and more crowding was observed.

In Table II, these observations are summarized for the undistributed portions of all the areas. Above M.T.L., the *Balanus* tended to overgrow the *Chthamalus*, whereas at the lower levels, undercutting was more common. This same trend was evident within each group of areas, undercutting being more prevalent on area 7 than on area 3, for example. The faster growth of *Balanus* at lower levels (Hatton 1938, Barnes and Powell 1953) may have resulted in more undercutting. When *Chthamalus* was completely covered by *Balanus* it was recorded as dead; even though death may not have occurred immediately, the buried barnacle was obviously not a functioning member of the population.

TABLE II. The causes of mortality of *Chthamalus stellatus* of the 1953 year group on the undisturbed portions of each area

Area no.	Height in ft from M.T.L.	No. at start	No. of deaths in the next year	Percentage of Deaths Resulting From:			
				Smothering by *Balanus*	Undercutting by *Balanus*	Other crowding by *Balanus*	Unknown causes
2	+3.5	28	1	0	0	0	100
3a	+2.2	111	81	61	6	10	23
3b	"	47	42	57	5	2	36
4	+1.4	34	14	21	14	0	65
5	+1.4	43	35	11	11	3	75
6	+1.0	27	11	9	0	0	91
7a	+0.7	42	38	21	16	53	10
7b	"	51	42	24	10	10	56
11a	0.0	21	13	54	8	0	38
11b	"	10	5	40	0	0	60
12a	0.0	60	57	19	33	7	41
12b	"	39	34	9	18	3	70
13a	−1.0	71	70	19	24	3	54
13b	"	69	62	18	8	3	71
14a	−2.5	22	21	24	42	10	24
14b	"	9	9	0	0	0	100
Total, 2–7	—	383	264	37	9	16	38
Total, 11–14	—	301	271	19	21	4	56

In Table II under the term "other crowding" have been placed all instances where *Chthamalus* were crushed laterally between 2 or more *Balanus*, or where *Chthamalus* disappeared in an interval during which a dense population of *Balanus* grew rapidly. For example, in area 7a the *Balanus*, which were at the high population density of 48 per cm^2, had no room to expand except upward and the barnacles very quickly grew into the form of tall cylinders or cones with the diameter of the opercular opening greater than that of the base. It was obvious that extreme crowding occurred under these circumstances, but the exact cause of the mortality of the *Chthamalus* caught in this crush was difficult to ascertain.

In comparing the survival curves of Figs. 2 and 3 within each area it is evident that *Chthamalus* kept free of *Balanus* survived better than those in the adjacent undisturbed areas on all but areas 2 and 14a. Area 2 was in the zone where adults of *Balanus* and *Chthamalus* were normally mixed; at this high level *Balanus* evidently has no influence on the survival of *Chthamalus*. On Stone 14a, the survival of *Chthamalus* without *Balanus* was much better until January when a starfish, *Asterias rubens* L., entered the cage and ate the barnacles.

Much variation occurred on the other 14 areas. When the *Chthamalus* growing without contact with *Balanus* are compared with those on the adjacent undisturbed portion of the area, the survival was very much better on 10 areas and moderately better on 4. In all areas, some *Chthamalus* in the undisturbed portions escaped severe crowding. Sometimes no *Balanus* happened to settle close to a *Chthamalus*, or sometimes those which did died soon after settlement. In some instances, *Chthamalus* which were being undercut by *Balanus* attached themselves to the *Balanus* and so survived. Some *Chthamalus* were partly covered by *Balanus* but still survived. It seems probable that in the 4 areas, nos. 4, 6, 11a, and 11b, where *Chthamalus* survived well in the presence of *Balanus*, a higher proportion of the *Chthamalus* escaped death in one of these ways.

The fate of very young *Chthamalus* which settled in the autumn of 1954 was followed in detail in 2 instances, on stone 15 and area 7b. The *Chthamalus* on stone 15 had settled in an irregular space surrounded by large *Balanus*. Most of the mortality occurred around the edges of the space as the *Balanus* undercut and lifted the small *Chthamalus* nearby. The following is a tabulation of all the deaths of young *Chthamalus* between Sept. 30, 1954 and Feb. 14, 1955, on Stone 15, with the associated situations:

Lifted by *Balanus*	: 29
Crushed by *Balanus*	: 4
Smothered by *Balanus* and *Chthamalus*	: 2
Crushed between *Balanus* and *Chthamalus*	: 1
Lifted by *Chthamalus*	: 1
Crushed between two other *Chthamalus*	: 1
Unknown	: 3

This list shows that crowding of newly settled *Chthamalus* by older *Balanus* in the autumn mainly takes the form of undercutting, rather than of smothering as was the case in the spring. The reason for this difference is probably that the *Chthamalus* are more firmly attached in the spring so that the fast growing young *Balanus* grow up over them when they make contact. In the autumn the reverse is the case, the *Balanus* being firmly attached, the *Chthamalus* weakly so.

Although the settlement of *Chthamalus* on Stone 15 in the autumn of 1954 was very dense, 32/cm², so that most of them were touching another, only 2 of the 41 deaths were caused by intraspecific crowding among the *Chthamalus*. This is in accord with the findings from the 1953 settlement of *Chthamalus*.

The mortality rates for the young *Chthamalus* on area 7b showed seasonal variations. Between October 10, 1954 and May 15, 1955 the relative mortality rate per day $\times$ 100 was 0.14 on the undisturbed area and 0.13 where *Balanus* had been removed. Over the next month, the rate increased to 1.49 on the undisturbed area and 0.22 where *Balanus* was absent. Thus the increase in mortality of young *Chthamalus* in late spring was also associated with the presence of *Balanus*.

Some of the stones transplanted from high to low levels in the spring of 1954 bore adult *Chthamalus*. On 3 stones, records were kept of the survival of these adults, which had settled in the autumn of 1952 or in previous years and were at least 20 months old at the start of the experiment. Their mortality is shown in Table III; it was always much greater when *Balanus* was not removed. On 2 of the 3 stones this mortality rate was almost as high as that of the younger group. These results suggest that any *Chthamalus* that managed to survive the competition for space with *Balanus* during the first year would probably be eliminated in the 2nd year.

Censuses of *Balanus* were not made on the experimental areas. However, on many other areas in the same stretch of shore the survival of *Balanus* was being studied during the same period (Connell 1961). In Table IV some mortality rates measured in that study are listed; the *Balanus* were members of the 1954 settlement at population densities and shore levels similar to those of the present study. The mortality rates of *Balanus* were about the same as those of *Chthamalus* in similar situations except at the highest level, area 1, where *Balanus* suffered much greater mortality than *Chthamalus*. Much of this mortality was caused by intraspecific crowding at all levels below area 1.

TABLE III. Comparison of the mortality rates of young and older *Chthamalus stellatus* on transplanted stones

Stone No.	Shore level	Treatment	Number of *Chthamalus* present in June, 1954		% mortality over one year (or for 6 months for 14a) of *Chthamalus*	
			1953 year group	1952 or older year groups	1953 year group	1952 or older year groups
13b	1.0 ft below MTL	*Balanus* removed	51	3	35	0
		Undisturbed	69	16	90	31
12a	MTL, in a tide pool, caged	*Balanus* removed	50	41	44	37
		Undisturbed	60	31	95	71
14a	2.5 ft below MTL, caged	*Balanus* removed	25	45	40	35
		Undisturbed	22	8	86	75

TABLE IV. Comparison of annual mortality rates of *Chthamalus stellatus* and *Balanus balanoides**

Area no.	Height in ft from M.T.L.	Population density: no./cm² June, 1954	% mortality in the next year
Chthamalus stellatus, autumn 1953 settlement			
1	+4.2	21	17
3a	+2.2	31	72
3b	"	29	89
6	+1.0	38	41
7a	+0.7	49	90
7b	"	52	82
11a	0.0	32	62
13e	−1.0	29	99
12a	(tide pool)	19	95
Balanus balanoides, spring 1954 settlement			
1 (top)	+4.2	21	99
1:Middle Cage 1	+2.1	85	92
1:Middle Cage 2	"	25	77
1:Low Cage 1	+1.5	26	88
Stone 1	−0.9	26	86
Stone 2	"	68	94

* Population density includes both species. The mortality rates of *Chthamalus* refer to those on the undisturbed portions of each area. The data and area designations for *Balanus* were taken from Connell (1961); the present area 1 is the same as that designated 1 (top) in that paper.

In the observations made at each census it appeared that *Balanus* was growing faster than *Chthamalus*. Measurements of growth rates of the 2 species were made from photographs of the areas taken in June and November, 1954. Barnacles growing free of contact with each other were measured; the results are given in Table V. The growth rate of *Balanus* was greater than that of *Chthamalus* in the experimental areas; this agrees with the findings of Hatton (1938) on the shore in France and of Barnes (1956a) for continual submergence on a raft at Millport.

TABLE V. Growth rates of *Chthamalus stellatus* and *Balanus balanoides*. Measurements were made of uncrowded individuals on photographs of areas 3a, 3b and 7b. Those of *Chthamalus* were made on the same individuals on both dates; of *Balanus*, representative samples were chosen

	CHTHAMALUS		BALANUS	
	No. measured	Average size, mm.	No. measured	Average size, mm.
June 11, 1954	25	2.49	39	1.87
November 3, 1954	25	4.24	27	4.83
Average size in the interval		3.36		3.35
Absolute growth rate per day x 100		1.21		2.04

After a year of crowding the average population densities of *Balanus* and *Chthamalus* remained in the same relative proportion as they had been at the start, since the mortality rates were about the same. However, because of its faster growth, *Balanus* occupied a relatively greater area and, presumably, possessed a greater biomass relative to that of *Chthamalus* after a year.

The faster growth of *Balanus* probably accounts for the manner in which *Chthamalus* were crowded by *Balanus*. It also accounts for the sinuosity of the survival curves of *Chthamalus* growing in contact with *Balanus*. The mortality rate of these *Chthamalus*, as indicated by the slope of the curves in Figs. 2 and 3, was greatest in summer, decreased in winter and increased again in spring. The survival curves of *Chthamalus* growing without contact with *Balanus* do not show these seasonal variations which, therefore, cannot be the result of the direct action of physical factors such as temperature, wave action or rain.

Seasonal variations in growth rate of *Balanus* correspond to these changes in mortality rate of *Chthamalus*. In Figure 4 the growth of *Balanus* throughout the year as studied on an intertidal panel at Millport by Barnes and Powell (1953), is compared to the survival of *Chthamalus* at about the same intertidal level in the present study. The increased mortality of *Chthamalus* was found to occur in the same seasons as the increases in the growth rate of *Balanus*. The correlation was tested using the Spearman rank correlation coefficient. The absolute increase in diameter of *Balanus* in each month, read from the curve of growth, was compared to the percentage mortality of *Chthamalus* in the same month. For the 13 months in which data for *Chthamalus* was available, the correlation was highly significant, $P = .01$.

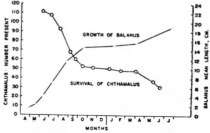

FIG. 4. A comparison of the seasonal changes in the growth of *Balanus balanoides* and in the survival of *Chthamalus stellatus* being crowded by *Balanus*. The growth of *Balanus* was that of panel 3, Barnes and Powell (1953), just above M.T.L. on Keppel Pier, Millport, during 1951-52. The *Chthamalus* were on area 3a of the present study, one-half mile south of Keppell Pier, during 1954-55.

From all these observations it appears that the poor survival of *Chthamalus* below M.H.W.N. is a result mainly of crowding by dense populations of faster growing *Balanus*.

At the end of the experiment in June, 1955, the surviving *Chthamalus* were collected from 5 of the areas. As shown in Table VI, the average size was greater in the *Chthamalus* which had grown free of contact with *Balanus*; in every case the difference was significant ($P < .01$, Mann-Whitney U. test, Siegel 1956). The survivors on the undisturbed areas were often misshapen, in some cases as a result of being lifted on to the side of an undercutting *Balanus*. Thus the smaller size of these barnacles may have been due to disturbances in the normal pattern of growth while they were being crowded.

These *Chthamalus* were examined for the presence of developing larvae in their mantle cavities. As shown in Table VI, in every area the proportion of the uncrowded *Chthamalus* with larvae was equal to or more often slightly greater than on the crowded areas. The reason for this may be related to the smaller size of the crowded *Chthamalus*. It is not due to separation, since *Chthamalus* can self-fertilize (Barnes and Crisp

TABLE VI. The effect of crowding on the size and presence of larvae in *Chthamalus stellatus*, collected in June, 1955

Area	Treatment	Level, feet above MTL	Number of Chthamalus	Diameter in mm Average	Diameter in mm Range	% of individuals which had larvae in mantle cavity
3a......	Undisturbed	2.2	18	3.5	2.7-4.6	61
"......	Balanus removed	"	50	4.1	3.0-5.5	65
4......	Undisturbed	1.4	16	2.3	1.8 3.2	81
"......	Balanus removed	"	37	3.7	2.5-5 1	100
5......	Undisturbed	1.4	7	3.3	2.8-3.7	70
"......	Balanus removed	"	13	4.0	3.5-4.5	100
6......	Undisturbed	1.0	13	2.8	2.1-3.9	100
"......	Balanus removed	"	14	4.1	3.0-5.2	100
7a & b..	Undisturbed	0.7	10	3.5	2.7-4.5	70
" ..	Balanus removed	"	23 ·	4.3	3.0-6.3	81

TABLE VII. The effect of predation by *Thais lapillus* on the annual mortality rate of *Chthamalus stellatus* in the experimental areas*

Area	Height in ft from M.T.L.	% mortality of Chthamalus over a year (The initial numbers are given in parentheses)					
		a: Protected from predation by a cage			b: Unprotected, open to predation		
		With Balanus	Without Balanus	Difference	With Balanus	Without Balanus	Difference
Area 3..	+2.2	73 (112)	25 (96)	48	89 (47)	6 (50)	83
Area 7..	+0.7	90 (42)	47 (40)	43	82 (51)	23 (47)	59
Area 11..	0	62 (21)	28 (18)	34	50 (10)	25 (16)	25
Area 12..	0†	100 (60)	53 (50)	47	87 (39)	59 (32)	28
Area 13..	−1.0	98 (72)	9 (77)	89	90 (69)	35 (51)	55

*The records for 12a extend over only 10 months; for purposes of comparison the mortality rate for 12a has been multiplied by 1.2.
†Tide pool.

1956). Moore (1935) and Barnes (1953) have shown that the number of larvae in an individual of *Balanus balanoides* increases with increase in volume of the parent. Comparison of the cube of the diameter, which is proportional to the volume, of *Chthamalus* with and without *Balanus* shows that the volume may be decreased to ¼ normal size when crowding occurs. Assuming that the relation between larval numbers and volume in *Chthamalus* is similar to that of *Balanus*, a decrease in both frequency of occurrence and abundance of larvae in *Chthamalus* results from competition with *Balanus*. Thus the process described in this paper satisfies both aspects of interspecific competition as defined by Elton and Miller (1954): "in which one species affects the population of another by a process of interference, i.e., by reducing the reproductive efficiency or increasing the mortality of its competitor."

The effect of predation by Thais

Cages which excluded *Thais* had been attached on 6 areas (indicated by the letter "a" following the number of the area). Area 14 was not included in the following analysis since many starfish were observed feeding on the barnacles at this level; one entered the cage in January, 1955, and ate most of the barnacles.

Thais were common in this locality, feeding on barnacles and mussels, and reaching average population densities of 200/m² below M.T.L. (Connell 1961). The mortality rates for *Chthamalus* in cages and on adjacent areas outside cages (indicated by the letter "b" after the number) are shown on Table VII.

If the mortality rates of *Chthamalus* growing without contact with *Balanus* are compared in and out of the cages, it can be seen that at the upper levels mortality is greater inside the cages, at lower levels greater outside. Densities of *Thais* tend to be greater at and below M.T.L. so that this trend in the mortality rates of *Chthamalus* may be ascribed to an increase in predation by *Thais* at lower levels.

Mortality of *Chthamalus* in the absence of *Balanus* was appreciably greater outside than inside the cage only on area 13. In the other 4 areas it seems evident that few *Chthamalus* were being eaten by *Thais*. In a concurrent study of the behavior of *Thais* in feeding on *Balanus balanoides*, it was found that *Thais* selected the larger individuals as prey (Connell 1961). Since *Balanus* after a few month's growth was usually larger than *Chthamalus*, it might be expected that *Thais* would feed on *Balanus* in preference to *Chthamalus*. In a later study (unpublished) made at Santa Barbara, California, *Thais emarginata* Deshayes were enclosed in cages on the shore with mixed populations of *Balanus glandula* Darwin and *Chthamalus fissus* Darwin. These species were each of the same size range as the corresponding species at Millport. It was found that *Thais emarginata* fed on *Balanus glandula* in preference to *Chthamalus fissus*.

As has been indicated, much of the mortality of *Chthamalus* growing naturally intermingled with *Balanus* was a result of direct crowding by *Balanus*. It therefore seemed reasonable to take the difference between the mortality rates of *Chthamalus* with and without *Balanus* as an index of the degree of competition between the species. This difference was calculated for each area and is included in Table VII. If these differences are compared between each pair of adjacent areas in and out of a cage, it appears that the difference and therefore the degree of competition, was greater outside the cages at the upper shore levels, and less outside the cages at the lower levels

Thus as predation increased at lower levels, the degree of competition decreased. This result would have been expected if *Thais* had fed upon *Balanus* in preference to *Chthamalus*. The general effect of predation by *Thais* seems to have been to lessen the interspecific competition below M.T.L.

Discussion

"Although animal communities appear qualitatively to be constructed as if competition were regulating their structure, even in the best studied cases there are nearly always difficulties and unexplored possibilities" (Hutchinson 1957).

In the present study direct observations at intervals showed that competition was occurring under natural conditions. In addition, the evidence is strong that the observed competition with *Balanus* was the principal factor determining the local distribution of *Chthamalus*. *Chthamalus* thrived at lower levels when it was not growing in contact with *Balanus*.

However, there remain unexplored possibilities. The elimination of *Chthamalus* requires a dense population of *Balanus*, yet the settlement of *Balanus* varied from year to year. At Millport, the settlement density of *Balanus balanoides* was measured for 9 years between 1944 and 1958 (Barnes 1956b, Connell 1961). Settlement was light in 2 years, 1946 and 1958. In the 3 seasons of *Balanus* settlement studied in detail, 1953-55, there was a vast oversupply of larvae ready for settlement. It thus seems probable that most of the *Chthamalus* which survived in a year of poor settlement of *Balanus* would be killed in competition with a normal settlement the following year. A succession of years with poor settlements of *Balanus* is a possible, but improbable occurrence at Millport, judging from the past record. A very light settlement is probably the result of a chance combination of unfavorable weather circumstances during the planktonic period (Barnes 1956b). Also, after a light settlement, survival on the shore is improved, owing principally to the reduction in intraspecific crowding (Connell 1961); this would tend to favor a normal settlement the following year, since barnacles are stimulated to settle by the presence of members of their own species already attached on the surface (Knight-Jones 1953).

The fate of those *Chthamalus* which had survived a year on the undisturbed areas is not known since the experiment ended at that time. It is probable, however, that most of them would have been eliminated within 6 months; the mortality rate had increased in the spring (Figs. 2 and 3), and these survivors were often misshapen and smaller than those which had not been crowded (Table VI). Adults on the transplanted stones had suffered high mortality in the previous year (Table III).

Another difficulty was that *Chthamalus* was rarely found to have settled below mid tide level at Millport. The reasons for this are unknown; it survived well if transplanted below this level, in the absence of *Balanus*. In other areas of the British Isles (in southwest England and Ireland, for example) it occurs below mid tide level.

The possibility that *Chthamalus* might affect *Balanus* deleteriously remains to be considered. It is unlikely that *Chthamalus* could cause much mortality of *Balanus* by direct crowding; its growth is much slower, and crowding between individuals of *Chthamalus* seldom resulted in death. A dense population of *Chthamalus* might deprive larvae of *Balanus* of space for settlement. Also, *Chthamalus* might feed on the planktonic larvae of *Balanus*; however, this would occur in March and April when both the sea water temperature and rate of cirral activity (presumably correlated with feeding activity), would be near their minima (Southward 1955).

The indication from the caging experiments that predation decreased interspecific competition suggests that the action of such additional factors tends to reduce the intensity of such interactions in natural conditions. An additional suggestion in this regard may be made concerning parasitism. Crisp (1960) found that the growth rate of *Balanus balanoides* was decreased if individuals were infected with the isopod parasite *Hemioniscus balani* (Spence Bate). In Britain this parasite has not been reported from *Chthamalus stellatus*. Thus if this parasite were present, both the growth rate of *Balanus*, and its ability to eliminate *Chthamalus* would be decreased, with a corresponding lessening of the degree of competition between the species.

The causes of zonation

The evidence presented in this paper indicates that the lower limit of the intertidal zone of *Chthamalus stellatus* at Millport was determined by interspecific competition for space with *Balanus balanoides*. *Balanus*, by virtue of its greater population density and faster growth, eliminated most of the *Chthamalus* by directing crowding.

At the upper limits of the zones of these species no interaction was observed. *Chthamalus* evidently can exist higher on the shore than *Balanus* mainly as a result of its greater tolerance to heat and/or desiccation.

The upper limits of most intertidal animals are probably determined by physical factors such as these. Since growth rates usually decrease with increasing height on the shore, it would be less likely that a sessile species occupying a higher zone could, by competition for space, prevent a lower one from extending upwards. Likewise, there has been, as far as the author is aware, no study made which shows that predation by land species determines the upper limit of an intertidal animal. In one of the most thorough of such studies, Drinnan (1957) indicated that intense predation by birds accounted for an annual mortality of 22% of cockles (*Cardium edule* L.) in sand flats where their total mortality was 74% per year.

In regard to the lower limits of an animal's zone, it is evident that physical factors may act directly to determine this boundary. For example, some active amphipods from the upper levels of sandy beaches die if kept submerged. However, evidence is accumulating that the lower limits of distribution of intertidal animals are determined mainly by biotic factors.

Connell (1961) found that the shorter length of life of *Balanus balanoides* at low shore levels could be accounted for by selective predation by *Thais lapillus* and increased intraspecific competition for space. The results of the experiments in the present study confirm the suggestions of other authors that lower limits may be due to interspecific competition for space. Knox (1954) suggested that competition determined the distribution of 2 species of barnacles in New Zealand. Endean, Kenny and Stephenson (1956) gave indirect evidence that competition with a colonial polychaete worm, (*Galeolaria*) may have determined the lower limit of a barnacle (*Tetraclita*) in Queensland, Australia. In turn the lower limit of *Galeolaria* appeared to be determined by competition with a tunicate, *Pyura*, or with dense algal mats.

With regard to the 2 species of barnacles in the present paper, some interesting observations have been made concerning changes in their abundance in Britain. Moore (1936) found that in southwest England in 1934, *Chthamalus stellatus* was most dense at M.H.W.N., decreasing in numbers toward M.T.L. while *Balanus balanoides* increased in numbers below M.H.W.N. At the same localities in 1951, Southward and Crisp (1954) found that *Balanus* had almost disappeared and that *Chthamalus* had increased both above and below M.H.W.N. *Chthamalus* had not reached the former densities of *Balanus* except at one locality, Brixham. After 1951, *Balanus* began to return in numbers, although by 1954 it had not reached the densities of 1934; *Chthamalus* had declined, but again not to its former densities (Southward and Crisp 1956).

Since *Chthamalus* increased in abundance at the lower levels vacated by *Balanus*, it may previously have been excluded by competition with *Balanus*. The growth rate of *Balanus* is greater than *Chthamalus* both north and south (Hatton 1938) of this location, so that *Balanus* would be likely to win in competition with *Chthamalus*. However, changes in other environmental factors such as temperature may have influenced the abundance of these species in a reciprocal manner. In its return to southwest England after 1951, the maximum density of settlement of *Balanus* was 12 per cm^2; competition of the degree observed at Millport would not be expected to occur at this density. At a higher population density, *Balanus* in southern England would probably eliminate *Chthamalus* at low shore levels in the same manner as it did at Millport.

In Loch Sween, on the Argyll Peninsula, Scotland, Lewis and Powell (1960) have described an unusual pattern of zonation of *Chthamalus stellatus*. On the outer coast of the Argyll Peninsula *Chthamalus* has a distribution similar to that at Millport. In the more sheltered waters of Loch Sween, however, *Chthamalus* occurs from above M.H.W.S. to about M.T.L., judging the distribution by its relationship to other organisms. *Balanus balanoides* is scarce above M.T.L. in Loch Sween, so that there appears to be no possibility of competition with *Chthamalus*, such as that occurring at Millport, between the levels of M.T.L. and M.H.W.N.

In Figure 5 an attempt has been made to summarize the distribution of adults and newly settled larvae in relation to the main factors which appear to determine this distribution. For *Balanus* the estimates were based on the findings of a previous study (Connell 1961); intraspecific competition was severe at the lower levels during the first year, after which predation increased in importance. With *Chthamalus*, it appears that avoidance of settlement or early mortality of those larvae which settled at levels below M.T.L., and elimination by competition with *Balanus* of those which settled between M.T.L. and M.H.W.N., were the principal causes for the absence of adults below M.H.W.N. at Millport. This distribution appears to be typical for much of western Scotland.

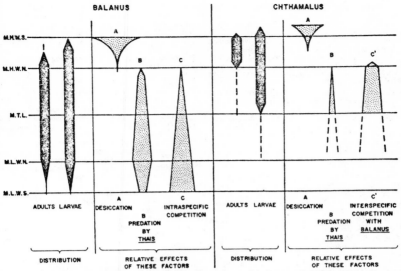

Fig. 5. The intertidal distribution of adults and newly settled larvae of *Balanus balanoides* and *Chthamalus stellatus* at Millport, with a diagrammatic representation of the relative effects of the principal limiting factors.

Summary

Adults of *Chthamalus stellatus* occur in the marine intertidal in a zone above that of another barnacle, *Balanus balanoides*. Young *Chthamalus* settle in the *Balanus* zone but evidently seldom survive, since few adults are found there.

The survival of *Chthamalus* which had settled at various levels in the *Balanus* zone was followed for a year by successive censuses of mapped individuals. Some *Chthamalus* were kept free of contact with *Balanus*. These survived very well at all intertidal levels, indicating that increased time of submergence was not the factor responsible for elimination of *Chthamalus* at low shore levels. Comparison of the survival of unprotected populations with others, protected by enclosure in cages from predation by the snail, *Thais lapillus*, showed that *Thais* was not greatly affecting the survival of *Chthamalus*.

Comparison of the survival of undisturbed populations of *Chthamalus* with those kept free of contact with *Balanus* indicated that *Balanus* could cause great mortality of *Chthamalus*. *Balanus* settled in greater population densities and grew faster than *Chthamalus*. Direct observations at each census showed that *Balanus* smothered, undercut, or crushed the *Chthamalus*; the greatest mortality of *Chthamalus* occurred during the seasons of most rapid growth of *Balanus*. Even older *Chthamalus* transplanted to low levels were killed by *Balanus* in this way. Predation by *Thais* tended to decrease the severity of this interspecific competition.

Survivors of *Chthamalus* after a year of crowding by *Balanus* were smaller than uncrowded ones. Since smaller barnacles produce fewer offspring, competition tended to reduce reproductive efficiency in addition to increasing mortality.

Mortality as a result of intraspecies competition for space between individuals of *Chthamalus* was only rarely observed.

The evidence of this and other studies indicates that the lower limit of distribution of intertidal organisms is mainly determined by the action of biotic factors such as competition for space or predation. The upper limit is probably more often set by physical factors.

References

Barnes, H. 1953. Size variations in the cyprids of some common barnacles. J. Mar. Biol. Ass. U. K. **32**: 297-304.

———. 1956a. The growth rate of *Chthamalus stellatus* (Poli). J. Mar. Biol. Ass. U. K. 35: 355-361.
———. 1956b. *Balanus balanoides* (L.) in the Firth of Clyde: The development and annual variation of the larval population, and the causative factors. J. Anim. Ecol. 25: 72-84.
——— and H. T. Powell. 1953. The growth of *Balanus balanoides* (L.) and *B. crenatus* Brug. under varying conditions of submersion. J. Mar. Biol. Ass. U. K. 32: 107-128.
——— and D. J. Crisp. 1956. Evidence of self-fertilization in certain species of barnacles. J. Mar. Biol. Ass. U. K. 35: 631-639.
Beauchamp, R. S. A. and P. Ullyott. 1932. Competitive relationships between certain species of freshwater Triclads. J. Ecol. 20: 200-208.
Clark, P. J. and F. C. Evans. 1954. Distance to nearest neighbor as a measure of spatial relationships in populations. Ecology 35: 445-453.
Clench, W. J. 1947. The genera *Purpura* and *Thais* in the western Atlantic. Johnsonia 2, No. 23: 61-92.
Connell, J. H. 1961. The effects of competition, predation by *Thais lapillus*, and other factors on natural populations of the barnacle, *Balanus balanoides*. Ecol. Mon. 31: 61-104.
Crisp, D. J. 1960. Factors influencing growth-rate in *Balanus balanoides*. J. Anim. Ecol. 29: 95-116.
Drinnan, R. E. 1957. The winter feeding of the oystercatcher (*Haematopus ostralegus*) on the edible cockle (*Cardium edule*). J. Anim. Ecol. 26: 441-469.
Elton, Charles and R. S. Miller. 1954. The ecological survey of animal communities: with a practical scheme of classifying habitats by structural characters. J. Ecol. 42: 460-496.
Endean, R., R. Kenny and W. Stephenson. 1956. The ecology and distribution of intertidal organisms on the rocky shores of the Queensland mainland. Aust. J. mar. freshw. Res. 7: 88-146.
Fischer, E. 1928. Sur la distribution géographique de quelques organismes de rocher, le long des cotes de la Manche. Trav. Lab. Mus. Hist. Nat. St.-Servan 2: 1-16.

Hatton, H. 1938. Essais de bionomie explicative sur quelques especes intercotidales d'algues et d'animaux. Ann. Inst. Oceànogr. Monaco 17: 241-348.
Holme, N. A. 1950. Population-dispersion in *Tellina tenuis* Da Costa. J. Mar. Biol. Ass. U. K. 29: 267-280.
Hutchinson, G. E. 1957. Concluding remarks. Cold Spring Harbor Symposium on Quant. Biol. 22: 415-427.
Knight-Jones, E. W. 1953. Laboratory experiments on gregariousness during setting in *Balanus balanoides* and other barnacles. J. Exp. Biol. 30: 584-598.
Knox, G. A. 1954. The intertidal flora and fauna of the Chatham Islands. Nature Lond. 174: 871-873.
Lack, D. 1954. The natural regulation of animal numbers. Oxford, Clarendon Press.
Lewis, J. R. and H. T. Powell. 1960. Aspects of the intertidal ecology of rocky shores in Argyll, Scotland. I. General description of the area. II. The distribution of *Chthamalus stellatus* and *Balanus balanoides* in Kintyre. Trans. Roy. Soc. Edin. 64: 45-100.
MacArthur, R. H. 1958. Population ecology of some warblers of northeastern coniferous forests. Ecology 39: 599-619.
Moore, H. B. 1935. The biology of *Balnus balanoides*. III. The soft parts. J. Mar. Biol. Ass. U. K. 20: 263-277.
———. 1936. The biology of *Balanus balanoides*. V. Distribution in the Plymouth area. J. Mar. Biol. Ass. U. K. 20: 701-716.
Siegel, S. 1956. Nonparametric statistics. New York, McGraw Hill.
Southward, A. J. 1955. On the behavior of barnacles. I. The relation of cirral and other activities to temperature. J. Mar. Biol. Ass. U. K. 34: 403-422.
——— and D. J. Crisp. 1954. Recent changes in the distribution of the intertidal barnacles *Chthamalus stellatus* Poli and *Balanus balanoides* L. in the British Isles. J. Anim. Ecol. 23: 163-177.
———. 1956. Fluctuations in the distribution and abundance of intertidal barnacles. J. Mar. Biol. Ass. U. K. 35: 211-229.

An Experimental Component Analysis of Population Processes[1]

By C. S. Holling

Perhaps the most important stimulus to studies of insect populations in Canada has been the development and application of the life-table approach (Morris 1962). This approach by itself can provide only a general description of events in a population. However, when life tables can be coupled with an understanding of processes such as predation, parasitism, and competition, that affect animal numbers, a clear insight into the causes of population fluctuations will be obtained. With this insight, control programs may be elevated from an *ad hoc* level. Hence there are pressing theoretical and practical reasons for applying the discipline of an analytical approach to a study of the operation of population processes. It is the aim of this paper to outline an approach that promises to provide this discipline by directing research along the causal pathways that mediate the action of each process.

The merit of this approach, and others with the same aim, can be judged by the degree to which the resulting structure meets three important requirements. First, population processes are so complex that they should be expressed in mathematical language, for man's mind, unaided, cannot appreciate the mode of action of all the components and their interactions. Second, it must *describe* real events accurately and consistently in order to permit useful predictions of future population outbreaks. But more is needed. To describe an event is obviously useful but it is even more useful to be able to explain the causes of the event. Hence the third and final requisite of the analysis is that it *explains* the mode of action of each process, so that the significant characteristics of a disease or parasite, for example, can be identified. This will suggest ways that populations can be manipulated through biological and silvicultural control programs and will identify those factors that should be measured to realize the full potential of the life-table approach. The fulfilment of these three requirements — to express mathematically, to describe, and to explain — might well represent an ideal situation impossible to attain. The benefits that can accrue, however, more than justify the attempt to approach this ideal.

The Experimental Component Analysis

The crux of the proposed approach lies in the belief that every process, however complex, can be fragmented into its constituent parts. A small group of these fragments is first isolated. Their effect, operating together, is then measured in a real or artificial situation that includes only these fragments. An explanation for their effect can then be postulated and, using standard scientific procedures, the validity of the postulates can be tested experimentally. When the postulates have been proved correct, additional fragments can be added successively, their action measured, hypotheses proposed, and their validity tested. In this manner a progressively more and more complex structure is built, and at each step a complete explanation is obtained. Moreover, the explanation, although first expressed in words, is ultimately expressed mathematically. This combination of experimentation and mathematical expression is mutually beneficial, for the experiments force reality on the mathematical expressions and the use of mathematics necessitates a precision that might not otherwise be sought.

[1]Contribution No. 805, Forest Entomology and Pathology Branch, Department of Forestry, Ottawa, Canada.

For this approach to be useful the process studied must be properly fragmented. Each process owes its characteristics to the effects of a number of variables, and it is these effects or components that are logically the significant fragments. If the process is at all complex, however, it is necessary to select a small group of them to provide a simple start for the analysis of the process. Clearly, the relative importance of components cannot provide a basis for selection since importance can only be appreciated after the analysis is complete. Some other criterion is needed which permits simplification without sacrificing reality. Even though different examples of a process might show a great deal of variety, an underlying similarity arises from certain components that are universally present. The diversity results from the addition of other components that occur in various combinations in each specific case. The former, universal components, can be called *basic* in that they underlie all manifestations of the process. The latter, occurring in some cases and not in others, can be termed *subsidiary*. Thus the initial group of components studied must be the basic ones, and the fragments progressively added will be the subsidiary components.

This approach can be conveniently termed an experimental component analysis, to distinguish it from other approaches that will be discussed later. A concrete example will better illustrate the advantages and the insight it provides.

To date I have analysed only one process, predation, in this manner. Since part of the analysis has already been published (Holling 1959a, b, 1961) and part will be published later, I shall include here only enough detail to demonstrate the use of the approach and to suggest the complexity of a typical population process.

The first step (Holling 1959a) was to identify as many as possible of the variables that affect predation and to group them into different functional categories. Data from the literature and general observations suggested five variables: density of the prey and predator and characteristics of the prey, the predator, and the environment. The universally-occurring variables proved to be prey and predator density and these gave rise to three basic responses. Thus there can be a functional response to prey density and a functional response to predator density such that the number of prey killed per predator changes as prey and predator densities change. These changes in consumption may promote a third response, a numerical response, in which the density of predators changes. The three subsidiary responses to changes in the characteristics of the prey, the predator, and the environment were shown to exert their effects through the three basic responses.

This study thus provided a verbal description of predation and suggested the various forms it could take. But it was only a description. To fulfil the need to express mathematically and to explain, it was then necessary to analyse precisely the mode of action of each of the basic responses. Only one of these, the functional response to prey density, has been studied so far, but the analysis has progressed far enough to provide considerable insight into the mechanics of attack by predators.

The component analysis was therefore continued by posing the same questions as before. That is, what components affect this basic response, the functional response, and which ones are basic and which ones subsidiary? In the course of observation and experimentation six components were identified. These are discussed elsewhere (Holling 1961) so that here we need only consider four of them: searching rate, time predator and prey are exposed, time spent handling prey, and hunger. It was felt that the first three are basic, in that every predator must search, must be exposed to prey, and must spend time handling prey.

Hunger, however, might not be universally important for there are certainly some predators that continue attacking prey even after they are satiated.

Again a situation was sought that included only the three basic components (Holling 1959b). Since it proved difficult to discover a real situation, an artificial one was devised. In this artificial attack, a blindfolded subject — the "predator" — searched for sandpaper discs — the "prey" — tacked to a table at different densities. The only components affecting the response were the rate of searching, the time available for all attack activities, and the time spent in handling discs after they were discovered, features common to every attack situation. As disc density rose, the number of discs picked up increased at a gradually decreasing rate until the curves levelled. A simple explanation was postulated and an equation devised. Independent tests proved the validity of these postulates and the resulting equation accurately described the response. The equation thus represents a basic functional response equation, for it incorporates the explanation of the operation of the three basic components.

Surprisingly, it also described the functional responses of insect parasites for which data had been published (Holling 1959b) and subsequently, the responses of three insect predators (unpublished data). It seems obvious, despite the descriptive power of the equation, however, that it does not incorporate a true and complete explanation of all these responses. Egg complement of the parasites and hunger of the predators must surely be important in some of the cases. The next logical step, therefore, was to add one more fragment, hunger, to the other three.

The responses of most invertebrate predators, and perhaps the lower vertebrate predators such as fish (Ivlev 1960) seem to owe their features to the operation of the three basic components plus hunger. The expansion of the equation to include the effects of hunger therefore represents a very important step, for it would provide a model for a wide array of attack situations. It is also the first point in the analysis in which a process of a high level of complexity is analysed. If the approach outlined is to be of any real value it must be capable of providing a precise explanation of this complex process. This analysis has just been completed, using praying mantids as subjects, and will form the basis of a paper to be prepared for later publication. For the present purpose, therefore, I shall only roughly and tentatively outline the explanation.

The earlier studies provide a sound basis for the present analysis. We already know that searching rate, time predator and prey are exposed, and handling time affect the response, and can postulate that hunger exerts its effect by modifying one or more of these three basic components. Searching rate, for example, might be high at low densities, when the predator is hungry, and low at high densities, when the predator is close to satiation. The analysis will be complete when the effects of hunger on the three basic components can be precisely stated.

Each of the basic components can be expanded into a number of subcomponents (Step 1, Fig. 1). It has been shown, for example, (Holling 1961) that searching rate is the product of the speed of movement of the predator relative to the prey, of the predator's distance of reaction, and of the ratio of successful captures to contacts. In a unit of time a moving predator of constant hunger sweeps an area equal in length to V and width to 2D. If it attacks every prey in this area, however, only a certain proportion will be captured (S). These three sub-components thus determine the magnitude of the searching rate. Similarly, the time a predator is exposed to prey can be expanded into at least two categories. Typically, during a 24 hour period of exposure only part of the

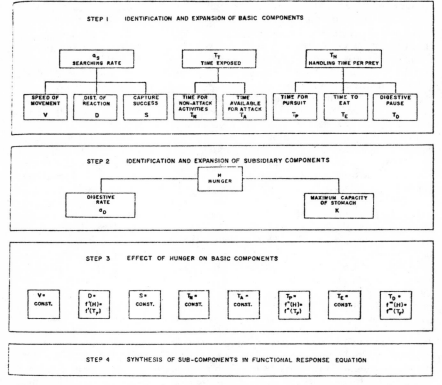

Fig. 1. Steps taken in the analysis of a functional response of predators to prey density.

24 hours is available for attack. If the predator operates solely in daylight then the time for attack (T_A) is limited to these hours. T_A is further restricted by the time taken for other non-attack activities (T_N), such as mating. The time spent handling prey, the last basic component, can also be expanded into at least three sub-components i.e. the time spent in pursuit of a located prey (T_P), the time spent eating a prey (T_E), and the time spent pausing to digest each prey consumed (T_D). It is specifically these eight sub-components that can conceivably be affected by hunger.

Before these effects can be explained it is necessary to describe hunger itself (Step 2, Fig. 1). Hunger, of course, is an internal motivational state, but it can be defined operationally in terms of its effect on behaviour. Now there are many behaviours it can affect but it is most useful to select one that most closely reflects the internal condition of the animal. A convenient measure is the weight of food necessary to return the animal to a condition of complete satiation, since this weight reflects the amount of food that has been evacuated and digested from the digestive tract. With increasing starvation in the praying mantid this measure of hunger rises at a decreasing rate until a sustained maximum is approached. A deductive equation that accurately describes this curve was devised. Its sub-components include rate of digestion and maximum amount of food the stomach will hold, and represent expansion of the hunger component.

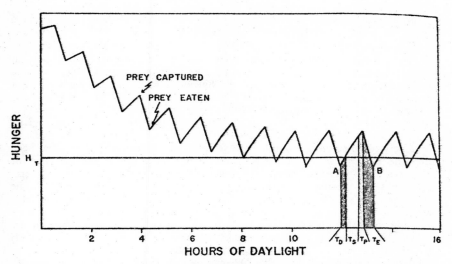

Fig. 2. Schematic representation of the changes in a predator's hunger during 16 hours of daylight. H_T = hunger level that just triggers attack; T_D = time spent in a digestive pause; T_S = time spent searching for prey; T_P = time spent in pursuit of prey; T_E = time spent in eating prey.

Having measured and described the characteristics of hunger, it then was possible to determine precisely how hunger affected the various sub-components of the three basic components (Step 3, Fig. 1). In the mantid only three were affected, the maximum distance of reaction (D), the time spent in pursuit of located prey (T_r), and the time spent in pausing to digest prey (T_D). As hunger increased, the distance of reaction and the time spent in pursuit increased, while the time spent digesting prey decreased. These three parameters were first expressed as functions of hunger, and then, with the aid of the hunger equation, as functions of time of food deprivation (T_r). These mathematical expressions therefore incorporated accurate descriptions and precise explanations of the effects of hunger.

In the final stage of the analysis these expressions were synthesized into a complete mathematical model of a functional response to prey density (Step 4, Fig. 1). This proved to be the most difficult task in all the analysis, and its solution can be better understood by consulting Fig. 2. This figure schematically portrays the changes in hunger of a predator at one prey density, during 16 hours of daylight, the time arbitrarily assigned as the time available for attack (T_A). The horizontal line represents the hunger threshold (H_T) above which the predator is hungry enough to attack and below which the predator is so satiated that it will not attack. The most striking feature is the historical, discontinuous character of the process. At the onset of daylight, after eight hours without food, the predator's hunger is fairly high and a certain time elapses before a prey is captured and eaten. The hunger is immediately lowered to a new value and then increases gradually until another prey is captured and eaten. This step-like process is continued until a stationary state is reached when the predator's hunger fluctuates between two fixed points, the lower occurring just after prey consumption and the higher just before. This stationary state persists to the end of the 16 hours of daylight. During the following night, when prey cannot be captured, the hunger would gradually rise to the level it achieved at the beginning, and at this point a new 16-hour period of daylight would begin. Any realistic model must preserve this historical and discontinuous nature of the attack process.

To retain this discontinuous character I felt that an approach using calculus was not warranted. Instead an algebraic approach was used which generated each cycle successively as time passed. Since the number of cycles or steps occurring over a given time equals the number of prey attacked at that density, the resulting difference equations can be expressed as a model of a functional response to prey density. Now the number of cycles generated during a given period depends upon the time taken to complete each. Consider the cycle in Fig. 2 between the points A and B, for example. At A, just after a prey has been eaten, the hunger is so low that it is below the hunger threshold and the predator will not attack. The time taken for the hunger to just reach the threshold therefore represents the digestive pause (T_D). Once the threshold is reached, the predator begins searching, and as it becomes hungrier it searches at a faster rate because of the influence of hunger on the distance of reaction. It continues searching long enough (T_s) to cover an area large enough to contain, on the average, somewhat more than one prey. The precise number above one is determined by the success the predator has in capturing a contacted prey (S). If the density of the prey is high then this area will be small, and the time taken to cover the area will be correspondingly low. Prey density, therefore, exerts its effect through T_s. Once a prey is located, a certain time is then taken to pursue it (T_r). When it is captured a final increment of time is added to permit consumption of the prey (T_E). The addition of these four time intervals equals the time taken for one cycle. Since mathematical expressions of these times are available, the total cycle time can be expressed mathematically. The resulting model, given a starting level of hunger, predicts the time taken for the first cycle as well as the new level of hunger that occurs at the end of the cycle. It is then possible to determine the time taken for the second cycle and each succeeding one until the time available for all attack activities elapses.

This final synthesis yields a model of a common type of functional response to prey density that is affected by the three basic components plus hunger. Future research will attempt to expand the model still further to include the effects of predator density, at which point additional analysis and experimentation will permit us to generate a numerical response, and thereby determine the results of the interaction of predator and prey. But even in its present incomplete stage the foregoing analysis suggests implications important for the development of population research.

Discussion
Models Produced by an Experimental Component Analysis

It will be clear by now, that predation is a process of high complexity, and we can expect that other population processes will be at least as complicated. Despite this complexity the experimental component analysis directs research in logical steps through the process, producing a concrete and realistic model of its mechanisms that fulfils the three requirements mentioned in the introduction, the need to express mathematically, to describe, and to explain.

The main problem that will be encountered in the analysis of other population processes will be in the final synthesizing stage. The way this problem was solved in the predation example, however, might well hold for other processes. Clearly, the acceptance of a solution should be based on how biologically real it is. Now the prime features of attack are its discontinuous, historical character, and many biological processes seem to operate in this way. Any synthesis must therefore preserve these features. This can be done, as in the predation example, by using an algebraic approach in which we express what happens in each step or cycle of the process. But it poses a problem. A calculus approach could be used and, even though it would be unreal, an approximate but incomplete description would result that would be relatively simple and manageable. An algebraic approach resulting in difference equations satisfies the need for reality but produces a most cumbersome expression, so cumbersome that a computer must be used to simulate the process. A most interesting feature appears, however, when a program for a digital computer is developed, for the program developed from the cumbersome algebraic expression is logical and straightforward. This occurs because the digital computer is designed to act in a step-like way, generating cycles consecutively. That is, the computer acts just as a predator does. This has far-reaching implications, for many biological phenomena have the same step-like, historical character. Hence we might well discover, as more biological processes are analysed in depth, that the explanations can best be expressed, not in the standard algebraic form, but as a program for a digital computer.

Potential Uses of the Models

The ultimate value of the experimental component analysis depends upon the usefulness of the resulting models. The model of the functional response to prey density provides an example of how useful these models can be. Since it closely mimics real attack situations it is possible to assign various sets of values to the parameters and so simulate the action of a variety of predators. This should show exactly the spectrum of situations we might find in nature and show very clearly which characteristics of a predator are most important in determining how many prey are destroyed. How important is size of predator and prey? What effect does speed of movement of the predator have on the efficiency of attack? Does a high rate of digestion of food confer a significant advantage to a predator, as a control agent? These, and similar questions, can be answered from the simulation studies. The procedure can also show the effects of different types of predation that differ in kind as well as degree. There are some predators, for example, that continue to attack and kill prey even though they are satiated. In such cases the hunger threshold is zero, and its exact effect can be determined using the model.

The simulation studies will have more than just theoretical interest, however. By directing attention to those features of predators that are most important in determining the number of prey destroyed, predators can be selected for biological control programs more efficiently. At the very least, we should be better able to predict what will happen if a predator is introduced to control a pest.

This practical use of the model can be achieved only because it meets the three requirements, to express mathematically, to describe, and to explain. But to meet these requirements simplicity has to be sacrificed. This poses a serious problem when we wish to use such models as the framework for a life-table study of a natural population. The models are so complex, an impossibly large number of parameters have to be measured. In the model of the functional response to prey density, for example, 14 different parameters and variables have to be measured; and this model concerns only one of three basic components of predation, which in turn is only one of a large number of processes that can affect animal numbers. 'Clearly, it would be impossible to measure so many factors in the field. Simpler models are required, the criterion of simplicity being that only four or five factors are necessary to use the model of any one process. There are two ways the model can be useful in this regard. First, since the exact importance of each parameter and variable is known, the model itself could be readily simplified to become useful in the field. Alternatively, the model, since it is a close approach to reality, can be used to test existing deductive models to see which ones describe most accurately and under what conditions they do so. It is possible, for example, that one deductive model might be best for one type of predation and another for another type. Either of these two methods will produce simple, descriptive models.

Other Approaches

There have been a number of approaches in population studies aimed at developing mathematical models. The considerable understanding we now have of the mode of action of predation, provides an excellent opportunity to assess the relative merits of these approaches and of the experimental component analysis.

Watt (1961) has recently published an excellent review of the relative merits of inductive and deductive models, so that it is sufficient here to assess very briefly their salient features. The inductive model simply represents the "best equation of an arbitrarily chosen simple type that accounts for the variation amongst an array of points" (Watt 1961, P. 27). It makes no pretence of presenting an insight into the operation of a process and, therefore, does not meet the need to explain the mode of action. It does describe the process but in such a limited way that the inductive model can be highly misleading. Typically, an inductive model can be developed to describe one example of a process, but once another example is tested a quite different model is necessary to describe it. Inductive models lack the ability to describe consistently.

Deductive models have had a prominent place in population biology. Unlike the inductive model, an explanation of the action of a process is incorporated in a set of reasonable assumptions that are expressed as one or more equations. But is it likely that the correct set of assumptions could ever be discovered without careful experimentation? In the functional response example there are four different stages in the development of the explanation (See Figure 1), and within each stage a quite elaborate series of assumptions would have to be made. Consider step 3 in Fig. 1, for example. There are eight parameters that could be affected by hunger and the mathematician would be faced with the problem of guessing exactly how each is affected. He would have to decide whether or not there are threshold effects, whether or not there are lag effects, whether or not the parameter increases with hunger up to asymptote, and whether or not the value of the parameter decreases when hunger exceeds a certain value. These four pairs of alternatives can yield 16 different possible equations for each of the eight parameters. To be conservative we might guess that 10 of these 16 possibilities are all equally reasonable, so that the probability of guessing the correct equation for any one parameter is 0.1. Since there are eight parameters the probability of choosing all of the eight correct equations is 10^{-8} i.e. one chance out of one million. Since this just concerns one of four stages it is clear that the explanation incorporated in a pure deductive model must inevitably be wrong, or, at the very least, incomplete. Moreover, it is highly unlikely it will even be descriptive. This would seem to explain why biologists express concern over the use of mathematics. They see that the many existing deductive models are unreal. The real parasite and the real predator acts in a quite different and much more complex way than the model parasite or predator. But this should not be an indictment of the use of mathematics in biology. Rather, it is an indictment of an approach which fails to meet the descriptive and explanatory requirements of a model.

Watt (1961) has significantly improved mathematical models by outlining an approach that yields combined deductive-inductive models. The logical basis of the approach is almost identical to that outlined here. The main difference is that fragmentation and experimentation is a more integral part of the experimental component analysis. This produces marked differences in the resulting models.

In practice, the deductive–inductive model is constructed by compiling a number of sets of reasonable assumptions each of which is incorporated in an equation. These equations are then fitted to published data to determine which one most closely describes the process. If none is descriptive, then new sets of assumptions are compiled until finally a consistently descriptive model is discovered. The strength of the approach is that it inevitably yields a model that accurately describes real events. Moreover, the deductive–inductive models proposed so far have accurately described a wide array of both laboratory and field situations. It is another question, however, whether the models incorporate the correct explanation of the mode of action of the process.

The experimental component analysis provides experimental evidence to demonstrate the action of each component separately. Thus a series of postulates is slowly constructed, and experimentally proved before the series is further expanded. In the deductive–inductive approach the mathematician relies on published data. Typically these data will concern only the end result of the action of a number of components so that a complete group of postulates has to be tested at once. As the process becomes more complex, it becomes less and less likely that the correct explanation will be incorporated in the set of assumptions, however descriptive they may be.

Two deductive–inductive models to describe attack by predators have been proposed independently by Watt (1959) and Rashevsky (1959). These are very similar, although Watt expanded his to include the effects of predator density and Rashevsky incorporated the effect of hunger in a more realistic way. Both describe very accurately a wide array of functional responses, and yet neither of them resemble the model arising from the experimental component analysis. Since the explanation incorporated in the latter model has been proved to be complete and correct, the explanation expressed by the two deductive–inductive models must either be wrong or must portray a different, and unreal causal chain that produces the same result as the real, more complex chain. The deductive–inductive models therefore do not seem to incorporate a true explanation of the action of a process. They admirably fulfil, however, the descriptive function required of a mathematical model. Moreover, they can be developed much more rapidly than a truly explanatory model. At the present stage of population biology they can therefore serve as a very powerful analytical tool that, at the least, makes predictions possible.

Conclusion

Three requirements were listed in the introduction that should be met to fully realize both theoretical and practical aims for population biology. Before the explanatory equations were developed for attack, however, I was not sure that it would be possible to meet all three requirements. The need to obtain precise mathematical equations, for example, might be incompatible with the need to explain the process. That is, the process might be so complex and so different in each situation that the explanation discovered would be so imprecise or specific that it would be futile to express it mathematically. It is now clear, however, that it is possible to meet all three requirements for predation and presumably for other processes as well.

The dynamics of natural populations are not only complex, they also often appear unique. There is nothing implicit in a mathematical approach, however, that requires one sweeping model that explains all situations. We can only expect that in every situation there will be common basic components, each of which belongs to one of a few types, so that their action can be expressed by one or a few general equations. An experimental component analysis hence has a very real advantage in that it directs research toward identifying and explaining these ubiquitous components. Certainly the interactions of these components and the kinds of subsidiary components in each population might be unique, but the prior identification and explanation of all possible components provides a general basis upon which to build an explanation for events in each population.

I feel, therefore, that the component analysis will provide a detailed insight into operations in natural populations. Whether this ultimately proves feasible or not, the potential benefits more than justify the attempt.

Summary

A clear insight into the operation of population processes can best be attained by developing mathematical expressions that not only fit observed data but also include an accurate and complete explanation of the action of the different factors involved. An approach is outlined that fulfils these requirements. As an example, a precise explanation of the action of the attack components of predation is given together with the method used to express this explanation in mathematical form.

This mathematical model describes and explains the attack process accurately, but only by sacrificing simplicity. If a useful degree of reality is to be retained complex processes require complex models. Fortunately, modern digital computers and the mathematical languages developed for them can readily handle this complexity. Because the models are realistic they promise to have a useful impact on practice and theory in population dynamics.

References

Holling, C. S. 1959a. The components of predation as revealed by a study of small mammal predation of the European pine sawfly. *Canadian Ent.* 91: 293-320.
Holling, C. S. 1959b. Some characteristics of simple types of predation and parasitism. *Canadian Ent.* 91: 385-398.
Holling, C. S. 1961. Principles of insect predation. *Ann. Rev. Entomology* 6: 163-182.
Ivlev, V. S. 1960. On the utilization of food by planktophage fishes. *Bull. Math. Biophysics* 22: 371-390.
Morris, R. F. (ed.) 1962. On the dynamics of spruce budworm populations. *Canadian Ent. Supplement* (in press).
Rashevsky, N. 1959. Some remarks on the mathematical theory of nutrition of fishes. *Bull. Math. Biophysics* 21: 161-183.
Watt, K. E. F. 1959. A mathematical model for the effect of densities of attacked and attacking species on the number attacked. *Canadian Ent.* 91: 129-144.
Watt, K. E. F. 1961. Mathematical models for use in insect pest control. *Canadian Ent. Supplement* 19: 1-62.

The Case for the Multispecies Ecological System, with Special Reference to Succession and Stability

G. Dennis Cooke, Robert J. Beyers, and Eugene P. Odum

While in a space capsule, man is a member of an ecosystem and, consequently, will be affected by all that goes on in this ecosystem. The success of long-term space flight may well depend on our success in developing a stable and long-lived life-support system. This paper will discuss some properties of ecosystems which promote stability and longevity and will propose that life-support systems must be developed within the conceptual framework of the mature multispecies ecosystem.

Several types of life-support systems have been designed or suggested to handle the following aspects of astronaut metabolism: gas exchange, food production, waste disposal, and nutrient and water regeneration. Of those proposed, only the storage system, designed for short flights, has been successfully tested. For flights of long duration, only bioregenerative systems appear to be feasible. The two-species gas exchange and/or food production model has received much attention. These have been described as the unialgal-man gas exchanger (ref. 1) and the *Hydrogenomonas*-man life-support system (fer. 2). In addition, Oswald et al. (ref. 3) have discussed the feasibility of an algae-bacteria-mammal system. One other type of life-support system has been suggested. This is the multispecies climax ecosystem, proposed by H. T. Odum (ref. 4), which will be the topic of this paper.

There are at least two approaches to the development of life-support systems. One of these consists of testing and later assembling separate biological, chemical, and mechanical components. The second consists of allowing groups of species known to occur together to reassemble and reorganize in a new environment into an integrated, self-maintaining system; this we call an ecosystem. Nature operates by the second method. Man uses the first in constructing his machines. We submit that the first is ecologically unsound and will prove to be unsuccessful. The balance of this paper will present evidence to show that the multispecies approach will provide the greatest opportunity for developing a successful long-term life-support system.

An ecosystem is any assemblage of organisms and their abiotic environment that has the following characteristics (refs. 5 and 6): structural organization, interdependency of components, homeostasis and regulation (external, internal, or both), limits and thresholds, and a development toward a steady state with increasing adaptation with and control of the physical environment (succession). There are four components of an ecosystem: (1) abiotic substances (organic and inorganic), (2) producers (autotrophs), (3) consumers (phagotrophs), which feed on larger particles, and (4) decomposers (osmotrophs), which derive their support from smaller or molecular size particles. Both of these latter categories decompose organic material and release products usable by producers.

One of the most important attributes of ecosystems is the unidirectional flow of energy from green plants through food webs to consumers and decomposers. The amount of photosynthate stored in excess of daytime community respiration is termed "net community photosynthesis." At night, part or all of this net storage is consumed by community respiration. A continued excess of community photosynthesis over community respiration leads to an accumulation of biomass. Eventually this accumulation of biomass stops as limits of light-input utilization are reached, or nutrients become limiting, or some physical threshold (such as space requirements) is reached.

Ecosystems have structure: biomass, stratification of both living and nonliving substances, and biochemical and species diversity. Depending on age and limitations of the physical environment, structure is more or less apparent in terms of the numbers of species. An ecosystem with many species per unit number of individuals may have a very complex food web as a result of niche (way of life) specialization by these species. Also, as the number of species increases, the number of homeostatic or regulatory mechanisms increases, and the organisms within and between the various trophic levels become more independent. There is increasing evidence to show that diverse ecosystems are also diverse biochemically (ref. 7).

The truly unique feature of ecosystems, however, is not structure or regulation, since these might be accomplished through external mechanical means, but the ability of ecosystems to develop, to come to a steady, self-maintaining, mature stage, often called a climax ecosystem. Some examples of mature systems adapted to particular physical factor regimes are temperate North American *Stipa-Biuteloua* perennial grassland, eastern deciduous oak-hickory forest, intertidal mangrove forest, tropical rain forest, and coral reef. For studies on mature natural ecosystems, see Golley et al. (ref. 8) and Odum and Odum (ref. 9). Oriental rice culture represents an agricultural system maintained by man which is more mature and stable than, for example, shifting row-crop tropical agriculture. The mature ecosystem tends to exhibit the maximum in structure and stability, within the limits imposed by the physical environment.

It is the mature ecosystem which we propose as the theoretical basis for the development of life-support systems. We believe that long-term stability, which is the result of the development of many homeostatic mechanisms through succession, must be the underlying concept in the development of a dependable life-support system. In other words, a multispecies system, with its associated high stability, has a far higher probability of survival than, for example, a two-species system.

A two-species system represents what we call "young nature." It has characteristics of early developmental stages of succession. A multispecies system, the result of a long developmental period, represents what we call "old nature" or a mature developmental stage (ref. 10).

A tabular model of succession (table I) has been prepared, with which we can compare the characteristics of old and young nature. Using some of the concepts of this model, we intend to compare two-species life-support systems and multispecies systems. We will show in greater detail some properties of ecosystems which demonstrate why we believe that a system with the characteristics of a mature ecosystem must be the basis of future life-support systems.

Not all of the ecosystem attributes of the model shown in table I are applicable, at present, to life-support systems. For example, no one envisions the possibility of man completing a life cycle in space. Other attributes are not well documented and need further research. Our remarks here will be confined primarily to attributes of ecosystem energetics and structure.

Many of the data which we will use to demonstrate the functional and structural events during succession have been obtained from the study of laboratory microcosms. These microecosystems are at least partially physically isolated from other ecosystems and in this respect are unnatural, since there is no export or import other than light and gas exchange with the atmosphere. However, the data obtained from these systems have particular applicability to the topic in question since a space capsule is a microecosystem.

The microcosm method has been described by Beyers (refs. 11 and 12). Materials from a natural ecosystem are brought into the laboratory and divided equally among a group of containers. Cross-seeding minimizes any possibility of divergence between microcosms. The systems are then placed on the desired regime of physico-chemical variables. Metabolism is measured by recording diurnal pH changes, and these data are translated into total CO_2 changes through the use of a graph depicting the relationship between microcosm pH and CO_2 changes. Microcosm biomass is determined by pouring an ecosystem into a tared weighing dish or through a tared millipore filter. The ma-

TABLE I.—*Model of Ecological Succession with Trends to be Expected in the Development of Ecosystems* [a]

	Ecosystem attributes	Developmental stages	Mature stages
Community energetics.	Gross production/community respiration	P/R ratio or one	Approaches one.
	Gross production/standing crop biomass	P/B ratio high	Low.
	Net community production (yield)	High	Low.
	Food chains	Linear, predominately grazing.	Web-like, predominately detritus.
Community structure.	Standing crop biomass & organic matter	Small	Large.
	Species diversity	Low	High.
	Biochemical diversity	Low	High.
	Stratification	Undeveloped	Well developed.
Life history.	Niche specialization	Broad	Narrow.
	Size of organism	Small	Large.
	Life cycles	Short, simple	Long, complex.
Nutrient cycling.	Free inorganic nutrients	Large	Small.
	Mineral cycles	Open	Closed.
	Nutrient exchange rate, organisms environment.	Rapid	Slow.
	Role animals in nutrient regeneration	Unimportant	Important.
Overall homeostasis.	Internal symbiosis (interdependence of organisms).	Low	High.
	Nutrient conservation	Poor	Good.
	Stability (resist external pertubation)	Poor	Good.
	Entropy	High	Low.
	Information	Low	High.

[a] Prepared by Eugene P. Odum.

terials are then dried and weighed. Similarly, total ecosystem chlorophyll is measured by filtering all or part of the system, reading the acetone extract at the appropriate wavelengths for the various pigments, and calculating the amount of chlorophyll according to Strickland and Parsons (ref. 13). These and other measurements are made at intervals during development or succession of the microcosm and, from these data, we have been able to show the course of some of the events of metabolic and structural succession.

In comparing the structure and function of old and young nature, we intend to emphasize these main points: (1) The two-species system represents young nature and has the advantage of a high rate of productivity per unit biomass, but with low stability. The multispecies system has a low photosynthesis-to-biomass ratio and must be large to support an astronaut, but it has the distinct advantage of multichannel stability. (2) The astronaut is part of a microecosystem, whether we are considering a two-species or a multispecies system, and therefore is part of the structure and function of the system. Depending on system stability, he will be more or less influenced by perturbations in its structure and function.

In figures 1 and 2, some data on succession in laboratory microcosms are plotted. In these experiments succession was initiated by inoculating material from a mature system into new medium. In young communities the rate of daytime photosynthesis exceeds that of night respiration, and biomass accumulates. Total or gross photosynthesis is high in the early stages. After about 70 days of succession, the ratio of daytime photosynthesis to night respiration ap-

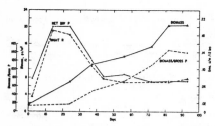

FIGURE 1.—Plots of net daytime photosynthesis, nighttime respiration, biomass and the ratio between biomass and gross or total photosynthesis against time in a microecosystem undergoing autotrophic succession. All data have been reduced to areal dimensions.

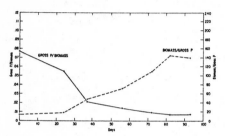

FIGURE 2.—Plots of the ratios of gross or total photosynthesis to biomass and vice versa against time in a microecosystem undergoing autotrophic succession. This figure illustrates the two types of efficiency outlined in the text.

proaches 1, and biomass reaches a stable value. At this point, the efficiency of the system is maximum, under a given set of environmental conditions, in that the highest level of biomass is maintained per unit of gross photosynthesis. Note also that the rates of respiration and photosynthesis are steady.

In the comparison of young and old nature, we are actually comparing two types of efficiency (fig. 2). In early developmental stages such as the two-species life-support system, the ratio of gross photosynthesis to biomass is very high—a small amount of structure is maintaining a high rate of photosynthesis. This is one type of efficiency, that which has been emphasized by proponents of the two-species system. If succession is allowed to proceed, whether by design or accident, the ratio drops. The trend in succession is to develop as large and diverse a structure, per unit of energy flow, as possible. Thus in early stages, the biomass-photosynthesis ratio is low; in a mature stage, the ratio is high. At climax a more complex structure, with a reduced waste of energy, allows the maintenance of the same biomass with a lower expenditure of energy or cost to the system (ref. 14). The more stable the system, both externally and internally, the less energy needed to maintain this biomass (ref. 15). In other words, as the system ages and develops structure, it becomes more efficient at maintaining that structure. This is the other type of efficiency, which we propose to be the basis of a stable life-support system.

The important point is that stability in these rates and ratios has been developed and will be maintained without external controls at maturity, while in young stages, stability must be constantly maintained through external controls. Mature microecosystems in the laboratory of Beyers have maintained themselves for years.

Another development during succession is the shift from an early planktonic-open water system to a detritus system in later stages. Recent studies (refs. 17–19) have shown that up to 90 percent or more of the metabolism of natural mature systems is in the detritus layer. The consequence of this in the use of a mature multispecies life-support system is that the astronaut must become a detritus-feeder, or a consumer of detritus-feeding organisms. This may, in fact, prove to be far more palatable than bacteria or algae since a great variety of vertebrates and invertebrates are detritus-feeders.

As communities develop, there is an increase in species diversity, and this has been assumed to contribute to stability (ref. 15). In a young ecosystem, there are a large number of unexploited ways of life, or niches. During succession, organisms from other ecosystems invade such structurally simple communities, or organisms that have been dormant or rare in the early stages become active and numerous. With time, the number of species per unit number of individuals, which we may call a species/number diversity index, increases. It is assumed on incomplete evidence that an increase in the diversity index favors the establishment of homeostasis in terms of checks and balances. During the early stages of succession, there may be "blooms" of the invader populations, often at the expense of one or more established species. Such blooms create perturbations that may seriously upset the balance within systems as does a "cancer" growth in individuals. For instance, in the development of unialgal life-support systems, Miller and Ward (ref. 20) have remarked on the difficulty of preventing the establishment of large populations of grazing zooplankton in their cultures. In a mature ecosystem, with most or all niches filled, the probability of blooms, or even the successful invasion by a new species, is very low (ref. 21); that is, the system now possesses stability. For example, the probability of invasion by extraterrestrial species, or a bloom of a mutated form of some component species, in a climax system would be much less than in an unsaturated system.

To summarize, the presence of many species not only means a diversity of energy pathways, but also the presence of a great many regulatory and symbiotic relationships. We cannot envision a two-species mechanical-bioregenerative system with this sophistication of control.

Another important trend from young to mature ecosystems, which is a direct result of increasing diversity, is the increase in complexity of food webs. In young stages, the number of species is small and, therefore, the number of pathways of energy transfer between producers and consumers is limited. In the two-species life-support system, this pathway is linear. In older stages, a great many species have had an opportunity to become established and the food web becomes more complex—so complex, in fact, that few have been completely described for any large natural area. Figures 3 and 4, based on the data of Paine (ref. 16), illustrate food webs of a simple and a complex ecosystem. The importance of food-web complexity to ecosystem stability is very apparent here. The top consumer in the more complex food web has the choice of 10 prey, and most of these prey also have several food chains from which to graze; whereas in the less complex system, the number of interactions is much lower. This is analogous to the backup systems built into the circuitry of space vehicles. In the microecosystem, the investigator has some choice about the number of species in the system, at least initially. During succession, several of the original groups of species may become extinct. For example, truly planktonic species do not survive succession to the climax stage.

Several examples will illustrate our point about the stability of mature ecosystems. In a climax forest, the outbreak of a pest is rare (ref. 21), but in a corn field (an ecosystem much like the two-species life-support system) large numbers of pests are common. In the forest a system of checks and balances ensures that an increase in insect population density is automatically followed by an increase in predator density. No such system exists in the corn field, and the farmer has to resort to pesticides or mechanical devices. With increasing insecticide resistance, he must resort to more and more potent chemicals. In practice, a combination of biological and physical control is usually optimum from man's standpoint. Our point is that we should fully utilize all possible self-regulation so as not to create unnecessary artificial substitutes.

Beyers (ref. 22) has shown that the metabolism of a complex climax ecosystem is con-

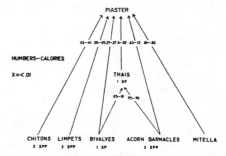

FIGURE 3.—The feeding relationships by numbers and calories of the Piaster-dominated food web at Mukkaw Bay, Washington. Piaster, N=1049; Thais=287, where N is the number of food items observed eaten by the predators. The specific composition of each predator's diet is given as a pair of fractions: numbers on the left, calories on the right (ref. 16).

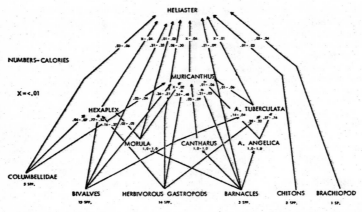

FIGURE 4.—The feeding relationships by numbers and calories of the Heliaster-dominated food web in the northern Gulf of California. Heliaster, N=2245; Muricanthus, N=113; Hexaplex, N=62; A. tuberculata, N=14; A. angelica, N=432; Morula, N=39; Cantharus, N=8. See figure 3 for further explanation (ref. 16).

siderably more independent of temperature than the metabolism of a simpler sewage community or a single organism. He postulated that the closer a living system approaches the integration of a balanced ecosystem, the less it is affected by temperature. This hypothesis may be expanded to state that the more complex the ecosystem, the less it will be affected by temperature extremes, adding stability to the system as a whole.

A mature ecosystem is also metabolically stable. Golueke and Oswald (ref. 23) have pointed out that the CO_2/O_2 exchange ratio of the plant used in setting up a biological gas exchanger must match that of the crew, since a very slight mismatch between human RQ and plant AQ would lead to the accumulation or loss of a certain fraction of the human oxygen demand and carbon-dioxide output per day. In the multispecies system, the RQ's of several kinds of heterotrophs would balance the AQ's of the several autotrophs so that a temporary imbalance with one species would be compensated for by other species.

Additional evidence of the increased stability for a multispecies system is shown by the data presented in figure 5. A climax microecosystem was irradiated with 10^6 rad in an acute dose. With the exception of the loss of one species (an ostracod), there was no visible effect on the system. However, when the system was used to initiate a new autotrophic succession, the results of radiation became apparent. As can be seen in figure 5, the rate of growth of the system was

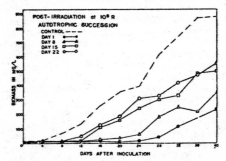

FIGURE 5.—Course of biomass increase with time in an autotrophic succession in a laboratory microecosystem irradiated at 10^6 rad. Successions were initiated by inoculating samples of the irradiated mature microecosystem at 1, 8, 15, and 22 days after irradiation. Control curve is from nonirradiated microecosystems.

decreased in comparison to the nonirradiated controls. However, this effect decreased with time, indicating the system's capacity for self-repair. The curve for each innoculation, made at weekly intervals after irradiation, shows progressive recovery. The greater the time after irradiation, the closer the curve approaches that of the controls. The principal primary producer and the organism accounting for the maximum biomass in this microcosm was a *Chlorella*. It is interesting to note that Posner and Sparrow (ref. 24) found that 90 percent of their pure culture *Chlorella* died after a dose of 23 000 rad. Our *Chlorella* showed no effects of radiation until they were irradiated at 2 000 000 rad and then took 40 days to die. These results may indicate that the system confers some radiation protection to its member organisms.

The interdependency of components in a climax microecosystem was clearly illustrated in a recent investigation in our laboratory by R. Gorden. Gorden has shown that a bacterium also present in the climax is an important source of thiamine, a requirement for the growth of the *Chlorella*. Close symbiosis between pairs of taxonomically unrelated species is an outstanding characteristic of the most successful natural communities, for example, lichen growths in the arctic or coral reefs in the tropics.

Although little information is presently available, it appears that during succession both the variety and amount of biochemicals increase (ref. 7). Many of these extrametabolites apparently have the properties of inhibitors or of growth promoters (ref. 25). These substances then are environmental hormones and act as regulators (ref. 7). Another biochemical change during succession is the increase in the quantity of accessory photosynthetic pigments, thus affording the ecosystem with more complete utilization of light as well as a complement of more stable pigments. The increased ability of mature systems to regulate themselves by internal chemical feedback means that less outside energy need be applied by man to achieve stability.

Beyers (refs. 11 and 26) has shown that there is a general pattern in the metabolism of aquatic ecosystems correlated with the onset of light and dark. The maximum metabolism occurs in the first half of the day or night period. The implications of this pattern for multispecies life-support systems have been discussed elsewhere (ref. 27). It must be admitted that there is the possibility of deleterious effects on an astronaut by this periodic reduction of photosynthesis and respiration. However, it does seem that the intensity of this phenomena decreases as the species diversity increases (ref. 11), adding another bit of evidence for the use of complex systems to support man in space.

Our main point in this discussion of ecosystem structure and function is high diversity and high stability in mature systems, low diversity and low stability in early stages. The more mature system has a builtin set of checks and balances which prevents internal disturbances and buffers the system against most external disturbances. Of course, no system is immune to severe perturbations and, in fact, the limits of ecosystem stability are strongly related to the stability of the physical environment (ref. 28).

There are certain distinct advantages to young nature, when viewed as life-support systems. These advantages are primarily energetic. That is, young ecosystems have high productivity rates per unit biomass, which means that they are more efficient gas exchangers in terms of O_2 produced or CO_2 absorbed per unit of biomass. As we have pointed out, however, there are serious disadvantages to young systems as well. It should now be apparent that, in the development of a life-support ecosystem for man, we must first select for system stability and longevity and then turn to the development of maximum productivity per unit biomass consonant with this stability. We cannot, as has been suggested by Miller and Ward (ref. 20), simply select organisms as needed on the basis of certain desirable functional characteristics and hope to integrate them successfully into a system. This attitude implies that the addition of new species will have no effect on resident species and this, of course, is ecologically unsound. It should also be apparent that homeostatic mechanisms of ecosystems are far more sophisticated and reliable than their mechanical counterparts. As H. T. Odum (ref. 4) has pointed out, man has yet to develop the miniaturization of circuitry that is found in ecosystems.

The area of a multispecies life-support ecosystem capable of supporting an astronaut has been estimated at 2 acres (ref. 4). This estimate is based on an expenditure of all but about 2 percent of the photosynthetic production on respiratory requirements of other components of the system. Obviously, practicality dictates some compromise between two-species mechanical systems and the multispecies system which depends on natural self-regulation.

To date, the major emphasis in the development of life-support systems has been on single components. These data are valuable and such work should be supported in the future. However, we believe, based on our knowledge of the properties of ecosystems, that future work must stress the development of a multispecies system.

Since such a system will necessarily be larger than proposed two-species systems, we need to determine which processes can be satisfactorily supplemented or replaced by mechanical or chemical devices. For example, the reduction of fecal material to small particles might best be handled by some mechanical method, thus eliminating the need for populations of consumers which ordinarily would fill this role. In other words, we may be able to reduce the predicted size of a multispecies life-support ecosystem without reducing the built-in stability of the system.

Finally, it is clear that a great deal of work must be directed toward an analysis of properties of ecosystems before we attempt to devise a multispecies life-support system. Part of the controversy regarding the simple vs complex system may be resolved by a set of experiments. Laboratory microecosystems ranging in complexity from a single alga and consumer to a highly complex system containing several representatives of each trophic level could be cultured under identical conditions. The climax systems and successional stages could be tested for metabolic and species stability under various stresses. Such stresses could include thermal manipulations, ionizing radiation, invasion by foreign species, and mechanical and photoperiodic stresses. We propose to perform such experiments in the future.

REFERENCES

1. MYERS, J. D.: The Use of Photosynthesis in a Closed Ecological System. *In:* Physics and Medicine of the Atmosphere and Space. John Wiley and Sons, Inc., New York, 1960, pp. 388–396.
2. BONGERS, L.; AND KOK, B.: Life-Support Systems for Space Missions. Develop. Ind. Microbiol., vol. 5, 1964, pp. 183–195.
3. OSWALD, W. T.; GOLUEKE, C. G.; AND HORNING, D. O.: Closed Ecological Systems. Jour. San Eng. Div., Proc. Amer. Soc. Civil Eng., vol. 91 (SA4), 1965, pp. 23–46.
4. ODUM, H. T.: Limits of Remote Ecosystems Containing Man. Amer. Biol. Teach., vol. 25, 1963, pp. 429–443.
5. ODUM, E. P.: Fundamentals of Ecology. W. B. Saunders Company, Philadelphia, 1959.
6. ODUM, E. P.: Ecology. Holt, Rinehart, and Winston, New York, 1963.
7. MARGALEF, R.: Successions of Populations. *In:* Adv. Frontiers of Plant Sci. edited by Raghu Vira. Instit. Adv. Sci. Culture, 1963.
8. GOLLEY, F.; ODUM, H. T.; AND WILSON, R. F.: The Structure and Metabolism of a Puerto Rican Red Mangrove Forest in May. Ecol., vol. 43, 1962, pp. 9–18.
9. ODUM, H. T.; AND ODUM, E. P.: Trophic Structure and Productivity of a Windward Coral Reef at Eniwetok Atoll, Marshall Islands. Ecol. Monogr., vol. 25, 1955, pp. 291–320.
10. ODUM, E. P.: Relationships Between Structure and Function in the Ecosystem. Jap. Jour. Ecol., vol. 12, 1962, pp. 108–118.
11. BEYERS, R. J.: The Metabolism of 12 Aquatic Laboratory Microcosms. Ecol. Monogr., vol. 33, 1963, pp. 281–306.
12. BEYERS, R. J.: The Microcosm Approach to Ecosystem Biology. Amer. Biol. Teach., vol. 26, 1964, pp. 491–498.
13. STRICKLAND, J. D. H.; AND PARSONS, T. R.: A Manual of Seawater Analysis. Fisheries Research Board of Canada Bull. No. 125, 1965.
14. MARGALEF, R.: On Certain Unifying Principles in Ecology. Amer. Nat., vol. 97, 1963, pp. 357–374.
15. CONNELL, J. H.; AND ORIAS, E.: The Ecological Regulation of Species Diversity. Amer. Nat., vol. 98, 1964, pp. 399–414.
16. PAINE, R. T.: Food Webb Complexity and Species Diversity. Amer. Nat., vol. 100, 1966, pp. 65–75.
17. ENGELMANN, M. D.: The Role of Soil Arthropods in the Energetics of an Old Field Community. Ecolog. Monogr., vol. 31, 1961, pp. 221–238.
18. MACFADYEN, A.: Metabolism of Soil Invertebrates in Relation to Soil Fertility. Ann. Appl. Biol., vol. 49, 1961, pp. 216–219.
19. ODUM, E. P.: Primary and Secondary Energy Flow in Relation to Ecosystem Structure. Proc. XVI Int. Cong. Zool., Aug. 20–27, 1963.
20. MILLER, R. L.; AND WARD, C. H.: Algal Bioregenerative Systems. USAF School of Aerospace Medicine SAM-TR-66-11, N66-27641, AD631191, 1966.
21. ELTON, C. S.: The Ecology of Invasions by Animals and Plants. Methuen Ltd.
22. BEYERS, R. J.: Relationship Between Temperature and the Metabolism of Experimental Ecosystems. Science, vol. 136, 1962, pp. 980–982.
23. GOLUEKE, C. G.; AND OSWALD, W. J.: Role of Plants in Closed Systems. Ann. Rev. Plant Physiol., vol. 15, 1964, pp. 387–408.
24. POSNER, H. B.; AND SPARROW, A. H.: Survival of *Chlorella* and *Chlamydomonas* After Acute and Chronic Gamma Radiation. Rad. Bot., vol. 4, 1964, pp. 253–257.
25. SAUNDERS, G. W.: Interrelations of Dissolved Organic Matter and Phytoplankton. Bot. Rev., vol. 23, 1957, pp. 389–409.
26. BEYERS, R. J.: The Pattern of Photosynthesis and Respiration in Laboratory Microecosystems. Mem. Ist. Ital. Idrobiol., vol. 18, 1965, pp. 61–74.
27. BEYERS, R. J.: The Microcosm Approach to Ecosystem Biology. Amer. Biol. Teach., vol. 26, 1964, pp. 491–498.
28. DUNBAR, M. J.: The Evolution of Stability in Marine Environments: Natural Selection at the Level of the Ecosystem. Amer. Nat., vol. 94, 1960, pp. 129–136.

COMMENTS

Dr. WARD. I would say that your suggestion is exactly that made previously by Jack Myers and later by myself and Dr. Miller. That is, a one-by-one selection of species and a study of their interactions where one knows what one has. Then build it up until you get your complexity from which you get your stability, rather than starting with a bucket and never knowing exactly what you have, where it is going, or what happens to it.

This last experiment you suggested is basically a rewording of the one-by-one selection to study what one can get away with towards life-support. I do not see that your suggestion is really divergent from what we have already suggested.

Dr. COOKE. This is not precisely what I am saying. Let me go back a way in the history of ecologic research and thought.

Most of the research in ecology has been done on simple single species or a pair of species. For example, one goes into the forest and studies the maple trees and some other fellow comes in and studies oak trees, or a variety of things. Then they hope to describe the ecosystem from what they have learned about each of these single species. As it turns out, the whole is a little greater than the sum of the parts. Consequently, you have to look at the whole system, as well.

We know from experience that, if you have an ecosystem which you have learned something about and have some data on, and then add a new species, you are going to have a new succession start, whether it be small or large. Our opinion is that, if we are going to try to do this kind of thing, let us put them together and learn how they interact. If one wants to learn specific processes in single organisms, fine. Perhaps we can get some physiologist to attack that, as well. It is the difference in the concept.

Dr. REPASKE. I believe there is a fundamental difference in principle between your ecosystem and what will occur in the space capsule. In your ecosystem, it is immaterial which species gains ascendency and which goes down.

In the capsule, we certainly are interested in one species maintaining itself, namely, the man. If you, in your ecosystem, would be interested in one of your flagellates or in one of your organisms, and had to do everything in order to have this maintain itself, then you would have an analogous system that we have in the capsule.

But, if you are just concerned about some life existing in some balance, and it is immaterial as to which species predominates, then I believe it is an entirely different system.

Dr. COOKE. I see your point. We, too, are interested in the astronaut surviving. That is our main point. If it takes a big system to support him, then it will have to be big. If he is one of many components in there, that is just the way you will have to look at it. We cannot sacrifice stability. This is the important thing.

Dr. REPASKE. I believe your approach is wrong, because, if you would propose your microecosystems from the standpoint of having one species and arbitrarily try to maintain this species in some constant number and let your other fluctuations occur as they do occur, then you have something analogous. If you let everything fluctuate randomly at will, you are comparing one system that has no basis for comparison with the other.

Dr. COOKE. I agree that these two systems are completely unalike. In other words, if we had a microecosystem in which we intended to maintain a fish in it, as a top consumer, we would have to make some efforts to see to it that everything that is needed for that fish is in there. This would be a pretty complex system in comparison to what we have. We have no vertebrates in it.

Dr. KRAUSS. I only wanted to remark that I believe there is nothing that you have said about the stability of ecosystems with which would be debated by any of us who have anything to do with ecology. But, we are dealing with a rather special situation here of a man in a space capsule. It is a pretty unstable situation, regardless of what the life-support situation is. But, because of trading off stability of an ecosystem to engineering, we are able to maintain just one species, as a matter of fact, for a reasonably long time.

We have had one up for the last 5 or 6 days, for instance. But you see, by introducing one or more microorganismal systems, we are, in fact, moving away from that single species to at least two or three, or something of this sort. The whole question in deciding about long-term life-support, I believe, is going to revolve about where you can trade off stability for engineering. A stable system is a tremendously expensive system from the point of view of space, weight, and energy consumption. A much more efficient system is a much more compact, compressed system, admittedly, with the dangers of instability. In this particular kind of situation, I think one has to invoke these basic principles of ecology with great caution. We are doing something which, in itself, is unstable, and is being maintained for a fairly limited period of time. With all of the resources of stability of the planet behind, it is compressed into an extremely efficient, functioning system for a relatively brief period.

Even a thousand days is a relatively brief period, but this thing comes out of a very broad-based stable ecology on the planet of which man is a part, an increasingly less stable part (as you well know), but still a part.

Dr. COOKE. Are you using the word "efficiency" a little differently than we are? We are talking about the efficiency of high production and small size.

Dr. KRAUSS. We have a specific mission here, you see. That is to get a man up for enough time to do whatever observational work he has to do and to get him back with the minimum of weight and volume to go with him during that period of time. Of course, we sacrifice much of the very sound and indisputable arguments of ecology in order to bring this about.

Dr. KOK. I still think there is no sacrifice whatsoever in the two-component system. The second point is right in your graphs. By including even three more components, your efficiency goes down with such dramatic numbers that in the very first curve you showed a factor of a 1000 in it. Not only is it a fallacy in stability conception, but also the efficiency is inhibited.

Dr. COOKE. I do not agree with saying there is a fallacy in stability. Would you rather ask whether an astronaut will come back or not?

All we are saying is that you have to invest some stability in the system, and engineering has not shown itself to be quite that good. How can you beat something that evolved for millions of years? That is pretty efficient engineering. You are dependent upon an ecosystem that is stable.

Dr. JENKINS. That is just the reverse. We have monocultures all over the world. Our whole agriculture is based on single species. Nobody tries to grow rice, peanuts and orange trees in the same soil, and we are talking about a system now where we have a few cubic meters, to support 8 or 10 men.

While your principles are correct, what relationship does it have to the spacecraft or even carrying it out, if you would extend this, to a lunar colony or to a Martian colony many years in the future? Would you ever get past the monoculture that we have on Earth?

Dr. COOKE. Your point is well made that obviously our agriculture is essentially a monoculture. Of course, we are sacrificing a lot of stability in our own largest earthly environment as a result of this monoculture. This is something else that ecologists are going to have to speak up about. We are getting more and more unstable. The temperature of the Earth is rising because our monoculture is attempting to replace fossil energy for solar energy through our own photosynthesis in our biosphere.

Dr. FOSTER. To put it more crudely, we want to bring them back in either case, but from an engineering viewpoint you are either going to scrub the trip completely in favor of stability or you are going for a thousand days, taking along with you as your major purpose, the best controller that has been so far invented, and that is man.

With certain instrumental aids, you have a man there who is operational and can control far better in a limited ecology than any other stable control that you can get.

Would you say that if there were going to be 10 or 20 species, this would be a reasonable estimate or would it be something else?

Dr. COOKE. I do not know.

Dr. TISCHER. If you were to set up a bacterial experiment at two levels, which is about as few as you can use to gain any information, you would have 2^{10} power experiments which would mean for one replication you would have to do a thousand determinations. If you double that, you have 2^{20} or a million determination, roughly. And so on and so on. It seems to me that added to the more simple cost factors involved is the cost of experimentation.

Dr. COOKE. I agree that it is an exremely complex problem. I can think of a lot of experiments, for example, that could easily answer some of your questions. It would be very nice, for example, to contain a small mammal, a rat or a mouse, or some other type organism, and attempt to build an isolated self-maintaining ecosystem. Such attempts have been made.

Dr. WARD. Do you consider Oswald's and Golneke's microecosystem as having considerably more stability than the other algal cultures, such as Dr. Krauss' which runs for weeks at a time? They had species diversity, in fact, quite a few. They had a pond of algae with associated bacteria that come out of an oxidation pond, plus the ones that inherently inhabit the gut of a mouse.

I do not see that any evidence to the fact that that particular system had turned out to be really any more stable than Jack Myers' balanced system, where he worked for several years with mice in algae cultures, or the stability of Dr. Krauss' algae cultures.

Dr. COOKE. We have yet to be impressed with a stable two-species ecosystem. I have not seen any evidence of it yet, if we want to call this an ecosystem. I have not seen a graph or paper that shows me a stable two-species life-support system that lasts more than a few hours to a few days.

Dr. Kok. Spaceflight is not interested in the ecosystem.

Dr. Cooke. A two-species system is an ecosystem. Have you shown me a system of life-support that has been stable for a long period of time?

Dr. DeCicco. I think we are confusing a completely closed system with a partially or mainly closed system. I do not recall anyone ever saying we could have a two-species, completely closed ecosystem. I think what we are trying to get at is as close to this as possible. We are exploring how complete a system we can get with two species and compare this to how complete a system we could get perhaps with more species. Then we look at the engineering problems involved in both and try to determine which would be the most advantageous.

Dr. Repaske. Do we really want an ecosystem or just a supplement to maintain a man? If the spacecraft were large enough, he could take corned beef sandwiches and oxygen, and we won't be concerned about an ecosystem at all. Because he probably cannot take enough tanks of oxygen and cannot take an adequate store of food, we simply want to supplement that which we cannot carry. I wonder if this necessarily involves us in an ecosystem or just a continuous supply of food and some exchange of atmosphere to maintain him?

Dr. Cooke. I think one shifts from the other, depending upon how long this voyage is going to be. I suppose the storage system could support a man maybe 30 or even 60 days. With complete storage of all the requirements in terms of tanks of gas and corned beef sandwiches, one gets past this point, and, sooner or later, you will have to have a more and more complete system for him.

Dr. Jenkins. The system has been designed for 700 days, four men, complete storage of food and oxygen. This is proposed for a Martian fly-by. However, when you get into 8 or 10 men and 1000 days, and some of these men landing on Mars, then it looks like the storage is getting out of hand. So a partial system of regeneration, at least for a partial regeneration of oxygen and water utilization, is needed. Actually, when you send up a man, you are sending up quite an ecosystem. There are many species of bacteria. In fact, if you didn't have a series of bacteria, the man probably would not have proper digestion.